Current Developments in Mathematics 2010

edited by

David Jerison
Barry Mazur
Tomasz Mrowka
Wilfried Schmid
Richard Stanley
Shing-Tung Yau

International Press
www.intlpress.com

Current Developments in Mathematics, 2010

2010 Mathematics Subject Classification. 00Bxx

ISBN 978-1-57146-228-2

Typeset using the LaTeX system.

Contents

Current Developments in Mathematics, 2010

The Arf-Kervaire problem in algebraic topology: Sketch of the proof

Michael A. Hill, Michael J. Hopkins, and Douglas C. Ravenel

ABSTRACT. We provide a sketch of the proof of the non-existence of Kervaire Invariant one manifolds using equivariant homotopy theory [**15**]. A treatment of the statement and history of the problem can be found in "The Arf-Kervaire problem in algebraic topology: Introduction". Our goal here is to introduce the reader to the techniques used to prove the result and to familiarize them with the kinds of computations needed.

CONTENTS

1. Introduction

The goal of this exposition is to provide a somewhat detailed sketch of the major points of our proof of the following theorem.

THEOREM 1.1. *Let $\theta_j \in \pi_{2^{j+1}-2}S^0$ be represented by a framed manifold of Kervaire invariant one. If θ_j is non-zero, then $j \leq 6$.*

For a detailed explanation of the statement of the theorem and the history of the problem, the reader is directed to look at our historical paper [**31**]: "The Arf-Kervaire problem in algebraic topology: Introduction". This paper will focus on explaining the technical and equivariant aspects of the proof. In particular, we shall review some of the salient equivariant homotopy, explore the norm functor, look at particular chain complexes, and compute differentials in the slice spectral sequence. The reader is also alerted that the determination of differentials via cohomology operations (§ 7.2) does not appear in the main paper; the proof here is novel. Several of the theorems (such as the Periodicity Theorem) are presented in slightly different generality as well.

A brief overview is as follows. We produce a C_8-equivariant spectrum $\tilde{\Omega}$ and prove a series of results from which the Kerviare proof follows immediately.

DETECTION THEOREM. *If θ_j exists, then the image in $\pi_*\tilde{\Omega}^{hC_8}$ is non-zero.*

GAP THEOREM. *The group $\pi_{-2}\tilde{\Omega}^{C_8}$ is zero.*

PERIODICITY THEOREM. *The homotopy groups of $\tilde{\Omega}^{hC_8}$ are 256-periodic.*

HOMOTOPY FIXED POINT THEOREM. *The spectra $\tilde{\Omega}^{C_8}$ and $\tilde{\Omega}^{hC_8}$ are weakly equivalent.*

Thus we begin with an equivariant spectrum $\tilde{\Omega}$, and we produce an ordinary spectrum by taking fixed points.

DEFINITION 1.2. Let Ω be the C_8 fixed points of $\tilde{\Omega}$:

$$\Omega = \tilde{\Omega}^{C_8}$$

The Homotopy Fixed Point Theorem shows that we could have equally well chosen to let Ω be the homotopy fixed points. We will sketch proofs of the Gap, Periodicity, and Homotopy Fixed Point theorems in this note. We shall not, however, prove the Detection Theorem, since this is an application of classical techniques of a somewhat different flavor than the rest. This has a slight expository difficulty related to which classes one must invert to produce our spectrum $\tilde{\Omega}$. We focus here on the close connection with the Lubin-Tate spectrum E_4, but even still, the flavor of § 8.2 is different. The interested reader is referred to the main paper for a proof of the Detection Theorem, the specifics of the inverted classes, and a more detailed treatment of the theorems sketched herein [**15**].

2. Basic introduction to equivariant homotopy

Though we will never make it explicit, we shall assume we are working in a rigid point-set model category whose homotopy theory is the usual homotopy theory of equivariant spectra [**20, 22, 23**]. In particular, we shall suppress any assumptions about fibrancy or cofibrancy of spectra in various constructions, and we shall focus on the most important aspects of the theory. The reader is suggested to consult the survey articles referenced here for more details about the general theory [**5, 24**].

There are many different (and inequivalent) notions of G-spectra. For our purposes, we consider "honest" G-spectra. These are G-spectra indexed by finite subspaces of an infinite dimensional inner product space that contains every irreducible representation of G infinitely often. The model categorical notions will not be explicitly spelled out, and instead we will focus on the computational aspects.

We begin with some standard notation. We will let $\mathcal{S}_G$ denote any of the nice categories of equivariant spectra. If X is a G-spectrum, then we shall also denote by X the image of the restriction functor to any subgroup H of G. The category of G-spectra is tensored and cotensored over G-spaces, and the function spectra objects are again G-spectra.

If X and Y are G-spectra, then we shall let $[X, Y]$ denote the homotopy classes of G-equivariant maps from X to Y. This is the same as π_0 of the G-fixed points of $F(X, Y)$. If we there is potential for confusion, we will adorn $[X, Y]$ with a subscript G to make clear the group. For a general G-space or spectrum X, we will let X^G denote the fixed point object.

We now recall that the forgetful functor $\mathcal{S}_G \to \mathcal{S}_H$ has a left and a right adjoint. The left adjoint, normally written $G_+ \wedge_H -$ is the "induction" functor, and we have a natural isomorphism

$$G_+ \wedge_H X \cong G/H_+ \wedge X$$

whenever X is a G-spectrum. The right adjoint, written $F_H(G, -)$ is the "coinduction" functor. One of the most delightful aspects of the theory is that for G a finite group, these are actually the same.

THEOREM 2.1 (Wirthmüller [**30**]). *We have a natural equivalence*

$$G_+ \wedge_H X \cong F_H(G, X).$$

We should think of this as the equivariant analogue of the statement that in spectra, finite coproducts ($G_+ \wedge_H X$) are the same as finite products ($F_H(G, X)$). For this reason, coinduction will not be talked about again.

This Wirthmüller isomorphism gives rise to natural isomorphisms

$$[G_+ \wedge_H X, Y]_G \cong [X, Y]_H \cong [X, F_H(G, Y)]_G \cong [X, G_+ \wedge_H Y]_G.$$

2.1. Spheres and cells. Given an orthogonal representation of V, we can form two spheres: $S(V)$, the unit sphere in V, and S^V, the one point compactification of V. These are related by a cofiber sequence in spaces

$$S(V)_+ \to S^0 \to S^V,$$

making S^V the unreduced suspension of $S(V)$. We remark that for every representation V, S^V has at least 2 fixed points: the origin and the point at infinity. The inclusion of these points will play a very important role in our proof.

For now, we fix some notation for the representations which will appear.

DEFINITION 2.2. Let σ denote the 1-dimensional sign representation of C_{2^n}.

Let $\lambda\colon C_{p^n} \to S^1$ denote the inclusion of the p^nth roots of unity.

Let $\lambda(k)$ denote the composite of λ with the kth power map on S^1.

Let ρ_{2^n} denote the regular representation of C_{2^n}, and let $\bar{\rho}_{2^n}$ denote the quotient by the trivial representation.

If we work localized at p, then $S^{\lambda(k)}$ is indistinguishable from $S^{\lambda(k')}$ whenever the p-adic valuation of k and k' agree. For this reason, we will restrict attention to the collection $\lambda(p^r)$. Since there is a σ for every cyclic 2-group, when we need to distinguish the sign representation for C_{2^k}, we will denote it by σ_k.

Although this gives us two very good notions of spheres, our notion of cells will be somewhat more elementary: we induce up from the standard notions of cells. Our n-cells are those G-spectra of the form $G_+ \wedge_H D^n$, where D^n is given the trivial H-action. Since D^n with the trivial H-action is visibly the restriction of D^n with the trivial G-action, we consider only cells of the form

$$G/H_+ \wedge D^n.$$

The boundary is of course $G/H_+ \wedge S^{n-1}$, and we build our notion of CW-complex with these as our cells. Additionally, understanding how to attach to X cells of the form $G/H_+ \wedge D^n$ amounts to understanding

$$[G/H_+ \wedge S^{n-1}, X]_G \cong [S^{n-1}, X]_H \cong \pi_{n-1}(X^H).$$

These kinds of homotopy groups form a rich structure and will be spelled out in more detail in § 4.

2.2. Geometric fixed points and isotropy separation. Perhaps the most difficult and in many ways counter-intuitive aspect of equivariant homotopy theory is the notion of "fixed points". In general, the fixed point functor from G-Spectra to ordinary spectral is very badly behaved. It is not symmetric monoidal, and it doesn't commute with infinite suspension. Even if the most elementary cases, understanding the fixed points is a non-trivial undertaking. For example, the fixed points for the sphere with a trivial G-action are shockingly complicated [**8, 7**].

THEOREM 2.3 (tom Dieck). *As rings, $\pi_0(S^{0^G})$ is the Burnside ring of finite G-sets $A(G)$.*

Already this is much more complicated that what we would expect the fixed points to be, namely S^0. This is also reflected in a splitting of the fixed points into a wedge of more complicated spectra.

Equivariant homotopy theory has machinery to correct this: geometric fixed points.

THEOREM 2.4. *There is a lax symmetric monoidal functor*

$$\Phi^G \colon \mathcal{S}_G \to \mathcal{S}$$

such that

$$\Phi^G(\Sigma^\infty X_+) = \Sigma^\infty X_+^G.$$

In other words, the geometric fixed point functor behaves exactly the way our intuition for fixed points suggests.

COROLLARY 2.5. *As rings, $\pi_0(\Phi^G(S^0)) = \mathbb{Z}$, and for a general virtual representation V, $\Phi^G(S^V) = S^{V^G}$.*

The definition of Φ^G will actually prove surprisingly useful. We begin with the notion of a classifying space and universal space for a family of subgroups of a group. Let $\mathcal{F}$ denote a family of subgroups of G such that if $H \in \mathcal{F}$, then all subgroups and conjugates of H are also in $\mathcal{F}$.

DEFINITION 2.6. A universal space associated to $\mathcal{F}$ is a G-space $E\mathcal{F}$ such that

$$(E\mathcal{F})^H \simeq \begin{cases} * & H \in \mathcal{F} \\ \emptyset & H \notin \mathcal{F}. \end{cases}$$

Thus the restriction of $E\mathcal{F}$ to any subgroup $H \in \mathcal{F}$ is H-equivariantly contractible. That such a space exists for a general group follows by mirroring Milnor's join construction of EG. For cyclic p-groups, we can find a distinguished representative $E\mathcal{F}$ for any family $\mathcal{F}$.

PROPOSITION 2.7. *Let $\mathcal{F}_m$ be the collection of proper subgroups of $C_{p^m} \subset C_{p^n}$, and let $E\mathcal{F}_m$ denote the usual universal space. Then*

$$E\mathcal{F}_m \simeq \lim_{\rightarrow} S(k\lambda(p^{m-1})),$$

the "unit sphere" in the infinite direct sum of copies of $\lambda(p^{m-1})$.

PROOF. If we restrict $\lambda(p^{m-1})$ to any proper subgroup of C_{p^m}, then we get the trivial representation. Similarly, the only fixed points for the C_{p^m} action is the origin. We therefore conclude that

$$S(k\lambda(p^{m-1}))^{C_{p^\ell}} = \begin{cases} S^{2k-1} & \ell < m \\ \emptyset & \ell \geq m. \end{cases}$$

The maps in the colimit are the standard inclusions $S^{2k-1} \to S^{2k+1}$, and so the colimit is contractible. □

Having built $E\mathcal{F}$, we can now build a much more important space: $\tilde{E}\mathcal{F}$. This is by definition the cofiber of the canonical map $E\mathcal{F}_+ \to S^0$, in spaces. As a consequence of the previous proposition, we can also identify a representative of the homotopy type $\tilde{E}\mathcal{F}_m$ for cyclic p-groups.

PROPOSITION 2.8. *A model for $\tilde{E}\mathcal{F}_m$ is given by*

$$\lim_{\rightarrow} S^{k\lambda(p^{m-1})}.$$

REMARK 2.9. In both of these propositions, the actual representations used don't matter. What mattered was that the fixed points of the proper subgroups become increasingly highly connected (which means we had an increasing number of copies of $\lambda(p^\ell)$ where $\ell \geq m-1$) and that there were no fixed points for C_{p^m} (which means that we have no copies of $\lambda(p^\ell)$ for $\ell \geq m$). We could, however, have added arbitrarily many copies of $\lambda(p^k)$ for $k < m$ and still produced the same homotopy type.

Though we will use several variants of $\tilde{E}\mathcal{F}$, by far the most important is when $\mathcal{F}$ is the family of all proper subgroups of G. In this case, $E\mathcal{F}_+ \to S^0$ is an equivalence when restricted to any proper subgroup, and $S^0 \to \tilde{E}\mathcal{F}$ is an equivalence on G-fixed points.

DEFINITION 2.10. Let $\mathcal{F}$ denote the family of proper subgroups of G. If X is a G-spectrum, then smashing X with the cofiber sequence defining $\tilde{E}\mathcal{F}$ yields the isotropy separation sequence:

$$E\mathcal{F}_+ \wedge X \to X \to \tilde{E}\mathcal{F} \wedge X.$$

The geometric fixed points of X are the G-fixed points of $\tilde{E}\mathcal{F} \wedge X$:

$$\Phi^G(X) = \left(\tilde{E}\mathcal{F} \wedge X\right)^G.$$

Both parts of this definition will be vitally useful. The isotropy separation sequence does exactly as the name implies: it separates X into parts that are built from cells of the form $G/G_+ \wedge D^n$ (the $\tilde{E}\mathcal{F} \wedge X$ part) and those that are built from induced cells (the $E\mathcal{F}_+ \wedge X$ part). Thus when we compute with this, the left most term is often handled by induction on the order of G (since all of these cells are induced). We can make this more precise. By the defining cofiber sequence for $\tilde{E}\mathcal{F}$, any map

$$G/H_+ \wedge Y \to \tilde{E}\mathcal{F} \wedge X$$

is automatically null-homotopic (the target is contractible when restricted to H).

Additionally, the map

$$E\mathcal{F}_+ \wedge X \to X$$

is a kind of generalized restriction, while $X \to \tilde{E}\mathcal{F} \wedge X$ is a localization. We can use this intuition to provide some heuristic justification for tom Diek's theorem. If we apply fixed points and take π_* then we have a long exact sequence (where we use $\Phi^G(S^0) = S^0$):

$$\cdots \to \pi_1(S^0) \to \pi_0\big((\Sigma^\infty E\mathcal{F}_+)^G\big) \to \pi_0({S^0}^G) \to \pi_0(S^0) \to 0.$$

If we can understand the homotopy of $(\Sigma^\infty E\mathcal{F}_+)^G$, then we can compute $\pi_0(S^0)$. However, since we are working stably, we have stable transfer maps in π_0 arising from the obvious map

$$G/H_+ \wedge X \to X.$$

In this context, this gives us a map

$$\pi_0({S^0}^H) = \pi_0(G/H_+ \wedge \Sigma^\infty E\mathcal{F}_+) \to \pi_0\big((\Sigma^\infty E\mathcal{F}_+)^G\big).$$

Induction on G then gives the result. We remark that though it is not shown here, all of the groups in question (with the exception of $\pi_1 S^0$) are actually finite rank free abelian groups. This can also essentially be shown by downward induction on the order of the group.

3. The norm and the construction of $MU_{(C_{2^n})}$

One of the most useful constructions in our main paper is that of the norm. This is a multiplicative induction that takes $\mathcal{S}_H$ to $\mathcal{S}_G$ whenever $H \subset G$, and it generalizes the Evens' transfer in group cohomology [**11**]. In the homotopical context, it was studied by Greenlees-May [**14**]. This section serves as a very general overview, giving intuition and some algebraic constructions that will help the reader understand the basic properties needed for the proof. In the main paper, a detailed construction with the proofs of the necessary equivariant properties is provided.

3.1. General properties of the norm. The basic idea of the norm is simple: induction in a category is essentially the coproduct of an object indexed by cosets. For representations, this is simply saying that we take the direct sum of $|G/H|$-copies of the representation and permute them in the obvious way, and for spectra, we instead use the wedge. In both of these examples, the underlying object is just the iterated coproduct, and the equivariance can be read out of the permutation action on G/H and the H-action on the underlying object.

Commutative algebras and E_∞-ring spectra have a different coproduct: the tensor or smash product. In these cases, again have a kind of induction, the norm. The two constructions are quite similar. We will sketch the one for commutative rings, and then state the salient properties of the one for spectra.

PROPOSITION 3.1. *There is a multiplicative functor $N_H^G\colon \mathcal{A}lg_H \to \mathcal{A}lg_G$ which is left adjoint to the forgetful functor i_H. The composite $i_H \circ N_H^G$ is the $|G/H|$-fold tensor power functor when H is a normal subgroup.*

In fact, the construction is obvious. The underlying algebra for $N_H^G(A)$ is

$$\bigotimes_{G/H} A.$$

To describe the equivariance, we chose a set of coset representative g_i (and we chose one of them to be the identity element). For any $g \in G$ and for all i, $gg_i = g_j h$ for some j and h. This allows us to specify the action: g acts on the factor indexed by g_i by sending it to the factor indexed by g_j and acting by h.

We can best understand the norm on the free commutative algebra on an H-set X, namely the polynomial algebra on X, $\mathbb{Z}[X]$.

PROPOSITION 3.2. *If X is an H-set, then*

$$N_H^G(\mathbb{Z}[X]) = \mathbb{Z}[G \times_H X].$$

This shows immediately the desired adjunctions. It also shows the failure of the restriction to simply be the tensor product of the underlying algebras. If H is not normal, then the restricting of $G \times_H X$ to H is not necessarily the disjoint union of copies of X.

It should be noted now that the norm functor actually lives as one on H-modules, not just H-algebras. On the H-algebras, it has a particularly nice description as a left adjoint. On the others, it's just the tensor power with the permutation G-action.

All of these statements are also true for commutative ring objects in H-spectra, namely the E_∞-ring spectra. Let $Comm_G$ denote the category of G-equivariant E_∞ ring spectra.

PROPOSITION 3.3. *The forgetful functor $i_H \colon Comm_G \to Comm_H$ has a left adjoint N_H^G, the underlying H-spectrum for which is the $|G/H|$-fold smash power when H is normal.*

The composite of the norm with the restriction as an endofunctor of $Comm_G$ is also understandable:

$$N_H^G(X) = G/H_+ \otimes X,$$

where we have used that E_∞ G-spectra are tensored over pointed G-sets.

Just as for commutative rings, there is a prolongation of this functor to all H-spectra. It will be essential to our proof of the Periodicity Theorem to have some computational results about it. These all rely on the following immediate fact.

PROPOSITION 3.4. *If W is a virtual representation of H, then $N_H^G S^W = S^{Ind_H^G W}$.*

This has the following very useful consequence. If $f \colon S^W \to X$ is an H-equivariant map, then we have a G-equivariant map

$$N_H^G(f) \colon S^{Ind_H^G W} \to N_H^G(X).$$

In particular, if X is the restriction of a G-equivariant E_∞-ring spectrum, then we can compose with the counit of the adjunction to get a new G-equivariant map

$$N_H^G(f) \colon S^{Ind_H^G W} \to X.$$

This gives us an additional, multiplicative way to promote H-equivariant homotopy classes to G-equivariant ones. Similarly, it is often quite simple to check that these are non-zero by restriction. If H is normal in G, then the restriction of $N_H^G(f)$ is essentially the $|G/H|$-fold power of f.

One of the most important properties of the norm is that it is well-behaved with respect to geometric fixed points. Since $S^0 = N_e^G S^0$, we expect terrible behavior for actual fixed points, and much of our work is spent boot-strapping up from the geometric fixed points.

THEOREM 3.5. *If X is an H-equivariant spectrum, then*

$$\Phi^G(N_H^G(X)) \simeq \Phi^H(X).$$

A very coarse sketch of the argument is as follows. We know from the earlier propositions that if $X = S^V$ and if $V_0 = V^H$, then

$$\Phi^G N_H^G(S^V) \simeq \Phi^G S^{Ind_H^G V} \simeq S^{V_0} \simeq \Phi^H S^V.$$

Now if we build X out of cells, then we have a diagram for building $N_H^G(X)$ out of cells. What is essential here is that the group G is acting on the diagram. The only non-induced cells in the diagram are those of the form $N_H^G(e)$, where e is a cell of X. Since geometric fixed points destroys induced cells, we are reduced to the case of spheres.

3.2. The spectrum $MU_{(C_{2^n})}$. A central role in our proof is played by the honest C_2-equivariant spectrum of Real bordism $MU_{\mathbb{R}}$ [**2, 19**]. This is the bordism spectrum of C_2-equivariant manifolds together with a Real structure on their stable normal bundle. This is a C_2-equivariant vector bundle where the fibers are complex vector spaces and where on fibers, the C_2-action is conjugate linear. With this in mind, the underling spectrum is just MU together with its complex conjugation action. If we take geometric fixed points (which corresponds heuristically to passage to fixed points for the Real manifolds), then we recover the unoriented bordism spectrum MO. These facts are central to our understanding of the spectrum.

Numerous authors have studied MU_R, computing its homotopy groups, showing that its fixed and homotopy fixed points agree, and determining its slices [**1, 17, 18**]. Additionally, we know that this is a C_2-equivariant E_∞-ring spectrum, so the norm will let us build spectra for any finite group G containing C_2 that have similar universal properties.

DEFINITION 3.6. Let $MU_{(G)}$ denote $N_{C_2}^G MU_{\mathbb{R}}$.

By Theorem 3.5, we know also that

$$\Phi^G MU_{(G)} = MO.$$

The spectrum $MU_{(C_{2^n})}$ is also an equivariant Thom spectrum. It carries geometric information of manifolds together with "2^{n-1}st roots of a Real structure". We will not exploit this geometry here, though a better understanding would give another proof of our Reduction theorem.

Just as in the ordinary case, $MU_{\mathbb{R}}$ carries orientations. Classically these are maps from MU, and in the C_2-equivariant case, these are maps from $MU_{\mathbb{R}}$. Many familiar spectra with natural C_2-actions have equivariant refinements, and their underlying complex orientation also refines to a Real one. In particular, K-theory becomes Atiyah's Real K-theory $K_{\mathbb{R}}$, and the standard orientation of K-theory becomes a real orientation of $K_{\mathbb{R}}$ [**3**]. Generalizing this, ongoing work of Avarett essentially shows that the push-forwards of the Lubin-Tate spectra E_n (endowed with their Hopkins-Miller C_2-action [**26**]) admit Real orientations [**4**]. This generalizes in the following way.

PROPOSITION 3.7. *If E is a G-equivariant E_∞ ring spectrum for G a finite group containing C_2, and if the restriction of E to C_2 admits a Real orientation*

$$u\colon MU_{(G)} \to E,$$

then there is a canonical equivariant E_∞ map

$$N(u)\colon MU_{(G)} \to E.$$

PROOF. This follows immediately from the adjunction

$$Map_{Comm_G}(MU_{(G)}, E) = Map_{Comm_{C_2}}(MU_{\mathbb{R}}, E).$$

Our desired map is the image of the orientation. $\square$

Thus $E_{2^{n-1}}$ admits a canonical map from $MU_{(C_{2^n})}$ that lifts the Real orientation.

3.3. Aside: Why not E_4? We pause here to explain part of the proof, namely why we do not use E_4. In fact, the Detection Theorem shows that the C_8-homotopy fixed points of E_4 is sufficient to detect the Kervaire classes. We know from the Nilpotence Theorem that there is a horizontal vanishing line in the homotopy fixed points spectral sequence [**6**], and that tells us immediately that the homotopy groups are periodic with period some power of 2. We can approach this by running the spectral sequence and seeing what classes on the zero line survive or by determining what sorts of bundles are orientable for $E_4^{hC_8}$ (which tells us immediately which classes must be permanent cycles).

However, even in the toy case of C_4 acting on E_2, it is very difficult to determine the differentials and extensions in the homotopy fixed point spectral sequence. Figure 1 shows the E_5 term of the spectral sequence (with an infinite number of copies of $\pi_* bo$ suppressed).

The spectral sequence terminates at E_{14}, and at that point, we can see that it is 32-periodic. What is not obvious from the E_2-term is that $\pi_{-2}E_2^{hC_4} = 0$. Determining the differentials on the class denoted by a star is one of the trickier parts of this computation. In fact, understanding the

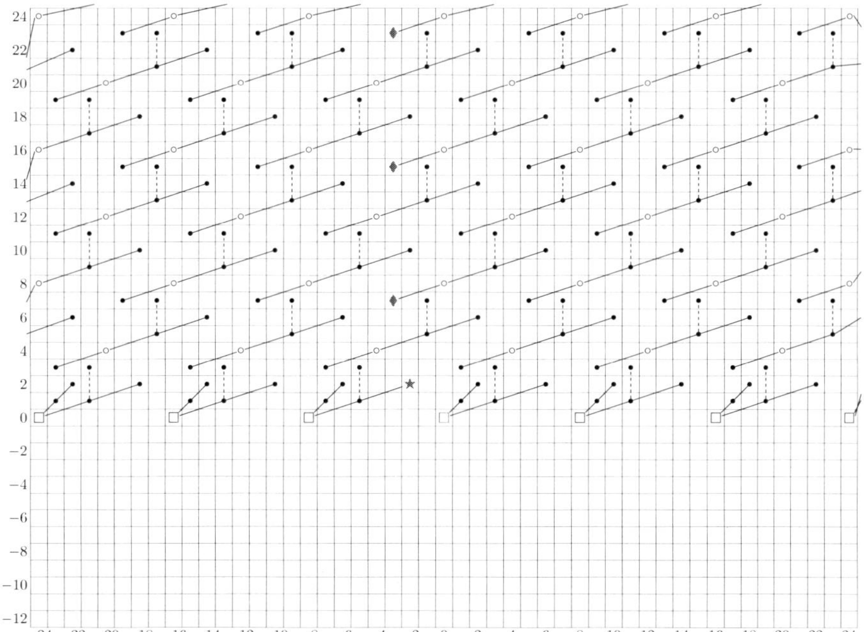

FIGURE 1. The E_5 term of the Homotopy Fixed Point Spectral Sequence for $E_2^{hC_4}$

analogue of the Kervaire differential required knowledge of some multiplicative relations in $\pi_{28}S^0$ at 2.

The C_8 case was visibly harder. While the C_4 case could be nicely presented at E_5 (in that each symbol represents either $\mathbb{Z}$ or a finite group), after much analysis with the slice spectral sequence, we believe that the C_8 case will not have a nice picture until E_{16}. Additionally, we know there will be a great many differentials before then, since there are fairly complicated patterns of differentials for C_2 and C_4 acting on E_4. It became clear early on that running the homotopy fixed point spectral sequence for $E_4^{C_8}$ would suffer from two major drawbacks:

(1) It would be very difficult to determine and to justify all differentials and extensions in the spectral sequence, and
(2) No one would ever believe it.

This led us to consider honest fixed points, rather than homotopy ones. While this might appear to be a Faustian bargain, trading rigidity for uncomputability, we had a major break-through with the creation of the slice spectral sequence. This provides a filtration of equivariant homotopy theory for a finite group G, and while the full form will not appear in this paper, this is the filtration of $MU_{(C_{2^n})}$ that makes computations doable. We should think of this as a much more tractible Atiyah-Hirzebruch spectral sequence. This filtration has a couple of antecedents: Dugger built the C_2-equivariant form and showed that $K_{\mathbb{R}}$ has the expected filtrations quotients and Hopkins-Morel, Voevodsky built an analogous filtration in the motivic context [**16, 27, 28, 29**].

The slice spectral sequence computes the honest fixed points of an equivariant spectrum (in fact, it does much more: since it is a purely equivariant filtration, it computes the equivariant homotopy groups as a Mackey functor). For spectra like E_2, E_4, and $MU_{(C_{2^n})}$, the E_2-term is actually easy to compute. In Figure 2, we show the E_5-page of the slice spectral sequence for C_4 acting on E_2.

Comparing with Figure 1 shows some obvious advantages. It is immediate that π_{-2} of the fixed points is zero, since it is zero on E_5. In fact, it is zero on E_2, and the Gap Theorem below shows that it is zero for a large family of equivariant spectra. The rigidity of equivariant homotopy theory helps us produce a large collection of permanent cycles in the fixed points, and these give us our desired periodicity.

The obvious advantage of the slice spectral sequence necessitates proving that the slices of E_4 have the required form. This in turn required first understanding the slices of $MU_{(C_{2^n})}$ and for various localizations thereof. It became clear that a judicious choice of localization of $MU_{(C_{2^n})}$ would sit as an intermediary between $MU_{(C_{2^n})}$ and E_4, being sufficiently strong to detect the Kervaire classes but also simple to work with. The reader is suggested to keep the spectrum E_4 with its natural C_8-action in mind, especially in § 8, where the intuition provided by E_4 is exactly correct.

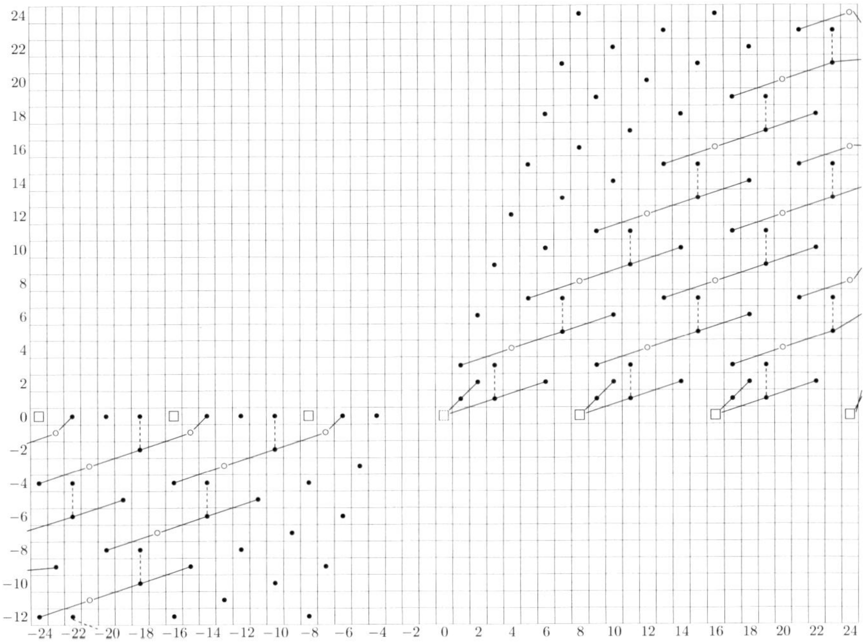

FIGURE 2. The slice E_5-term for C_4 acting on E_2

4. Mackey functors and $RO(G)$-Graded homotopy

4.1. Introduction to Mackey functors. One of the biggest hurdles facing homotopy theorists interested in equivariant algebraic topology is the more complicated notion of homotopy groups. Classically, the homotopy groups of a spectrum are simply abelian groups. In the equivariant context, the homotopy groups are objects in a more complicated abelian category, the category of Mackey functors [**9**]. The complexity, and the definition of Mackey functors, arises from the larger collection of "points" in the equivariant context. In addition to $* = G/G$, we have any coset space G/H. These G-sets are linked by a large collection of stable G-maps, and these provide a collection of maps between the homotopy groups.

DEFINITION 4.1. Let $\mathcal{O}_G$ denote the category of finite G-sets.

DEFINITION 4.2. A Mackey functor $\underline{M}$ is a pair of functors $M^*\colon \mathcal{O}_G^{op} \to \mathcal{A}b$ and $M_*\colon \mathcal{O}_G \to \mathcal{A}b$ which send disjoint unions to direct sums, such that

$$M^*(X) = M_*(X) =: \underline{M}(X)$$

for all objects X, and such that if

$$\begin{array}{ccc} X & \xrightarrow{f} & Y \\ {\scriptstyle g}\downarrow & & \downarrow{\scriptstyle h} \\ Z & \xrightarrow[k]{} & W \end{array}$$

is a pull-back square in G-sets, then

$$\begin{array}{ccc} \underline{M}(X) & \xrightarrow{M_*(f)} & \underline{M}(Y) \\ \uparrow{\scriptstyle M^*(g)} & & \uparrow{\scriptstyle M^*(h)} \\ \underline{M}(Z) & \xrightarrow[M_*(k)]{} & \underline{M}(W) \end{array}$$

is commutative.

We will denote $M^*(f)$ by f^* and $M_*(f)$ by f_*, and we will call f^* a restriction and f_* a transfer. If $f\colon G/H \to G/G$ is the obvious map, then following this we will often write Tr_H^G for f_* and Res_G^H for f^*. This terminology will be made somewhat more clear by connecting these to standard constructions in homotopy theory.

DEFINITION 4.3. If X is a G-spectrum, then the kth homotopy group of G is the Mackey functor

$$\underline{\pi_k(X)}(B) = [B_+ \wedge S^k, X]_G.$$

When $B = G/H$, we will also let

$$\pi_k^H(X) = \underline{\pi_k(X)}(B).$$

This description makes the restriction maps immediate: they arise by precomposition. For the transfer maps, we can work one of two ways.

(1) In the equivariant stable homotopy category, associated to a map $\Sigma^\infty B_+ \to \Sigma^\infty C_+$ there is a transfer map $\Sigma^\infty C_+ \to \Sigma^\infty B_+$. Precomposing with these transfer maps gives the desired transfer maps.
(2) Using the isomorphisms

$$[G/H_+ \wedge X, Y]_G \cong [X, Y]_H \cong [X, G/H_+ \wedge Y]_G,$$

we can conclude that there is a natural isomorphism

$$[B_+ \wedge X, Y] \cong [X, B_+ \wedge Y]$$

for all G-sets B and for all X and Y. We can therefore consider the compositions in the target, rather than in the source, to get our transfer maps.

Since the object function is assumed to be additive, it suffices to understand what happens on the orbits G/H:

$$\pi_k^H(X) = [G/H_+ \wedge S^k, X]_G = [G_+ \wedge_H S^k, X]_G = [S^k, X]_H = \pi_k(X^H).$$

Thus the value of the homotopy group Mackey functor on G/H_+ is exactly π_k of the H-fixed points of X. Moreover, if $H \subset K \subset G$, then we have a G-map $G/H \to G/K$ which gives us a map in the other way on the Mackey functor homotopy groups. This coincides with the restriction map $\pi_k(X^K) \to \pi_k(X^H)$ induced by inclusion of fixed points. Similarly, the transfer map is essentially "summing over the cosets".

Since they are a diagram category of abelian groups, Mackey functors form an abelian category. This abelian category is closely tied to the equivariant stable homotopy category.

PROPOSITION 4.4. *Let $\mathcal{S}_{G,0}$ denote the full subcategory of $\mathcal{S}_G$ consisting of those objects X such that $\underline{\pi}_i(X) = 0$ for $i \neq 0$. Then $\underline{\pi}_0$ induces an equivalence of categories between the homotopy category of $\mathcal{S}_{G,0}$ and the category of Mackey functors.*

In particular, for any Mackey functor $\underline{M}$, we have an associated Eilenberg-Mac Lane spectrum $H\underline{M}$ which satisfies

$$\underline{\pi_k(H\underline{M})} = \begin{cases} \underline{M} & k = 0, \\ 0 & \text{otherwise.} \end{cases}$$

We will be most concerned with $H\underline{\mathbb{Z}}$.

Just as with abelian groups, we can define "bilinear" pairings of two Mackey functors $\underline{M}$ and $\underline{N}$ into a third, $\underline{P}$. This is a natural transformation $M^* \otimes N^* \to P^*$:

$$\langle -, - \rangle_X \colon \underline{M}(X) \otimes \underline{N}(X) \to \underline{P}(X)$$

together with the condition that for any $f\colon X \to Y$, the following diagram commutes

$$\begin{array}{ccccc} \underline{M}(X) \otimes \underline{N}(Y) & \xrightarrow{\underline{M}(X)\otimes f^*} & \underline{M}(X) \otimes \underline{N}(X) & \xrightarrow{\langle -,- \rangle_X} & \underline{P}(X) . \\ {\scriptstyle f_*\otimes \underline{N}(Y)}\downarrow & & & & \downarrow{\scriptstyle f_*} \\ \underline{M}(Y) \otimes \underline{N}(Y) & & \xrightarrow[\langle -,- \rangle_Y]{\qquad\qquad\qquad} & & \underline{P}(Y) \end{array}$$

Put more sucinctly:

$$\langle Tr(x), y \rangle_Y = Tr(\langle x, Res(y) \rangle_X).$$

A Green functor is a Mackey functor $\underline{M}$ together with a bilinear pairing $\underline{M} \otimes \underline{M} \to \underline{M}$. Since this definition is opaque, we will unpack it slightly. A Green functor $\underline{M}$ is one for which $\underline{M}(X)$ is a ring for all X and for which all restriction maps f^* are ring homomorphisms. This means that for any $f\colon X \to Y$, $\underline{M}(X)$ is a $\underline{M}(Y)$-module via f^*, and the condition for the pairing can be restated as follows: f_* is a map of $\underline{M}(Y)$-modules. It is then immediate how one defines modules over a Green functor.

The most important Green functor is the Burnside ring $\underline{A}$. This associates to a finite G-set X the Grothendieck group of finite G-sets over A, and restriction and transfer are just pull-back and composition. The pullback of two G-sets over X gives a product on this set, making $\underline{A}(X)$ into a ring for all X. It is then trivial to check that transfers are module maps over the restrictions. It is immediate that $\underline{A}(G/G) = A(G)$, the Burnside ring of finite G-sets. If H is a subgroup of G, then the map

$$\big(f\colon X \to G/H\big) \mapsto (f^{-1}(eH) \to eH)$$

identifies $\underline{A}(G/H)$ with $\underline{A}(H)$. In particular, $\underline{A}(G/\{e\}) = \mathbb{Z}$.

The Burnside Mackey functor plays the role for Mackey functors that $\mathbb{Z}$ plays for abelian groups. In particular, every Mackey functor is a module over the Burnside ring Mackey functor. The module structure is actually very easy to spell out: transfers are module maps. If we look at a particular orbit G/H, then this is the transfer, in the Burnside ring, of the element $1 \in A(H)$. By the defining property of a pairing of Mackey functors, we then know how it acts on an element $x \in \underline{M}(G/G)$:

$$[G/H] \cdot x = Tr_H^G Res_G^H(x).$$

A more general statement is also easy to form, using exactly this. A general element in $\underline{A}(X)$ is a G-set over X: $f\colon Y \to X$. The above observation shows us how to pair $\underline{A}(X)$ and $\underline{M}(X)$:

$$f \otimes x \mapsto f_* f^*(x).$$

Understanding this action will help us make several computations related to equivariant homotopy groups.

We have a few distinguished ideals in the Burnside ring. First, note that if a G-set is induced, then any G-set we get by pull-back is also induced. This says that there is a "transfer" ideal in the Burnside ring. We can push slightly farther, though. If $\mathcal{F}$ is a family of subgroups of G, then we can form an ideal in the Burnside ring associated to $\mathcal{F}$, $\underline{I_{\mathcal{F}}}$ as follows: begin with the ideal of $A(G)$ generated by $[G/H]$ for all $H \in \mathcal{F}$, and then close up under restrictions and transfers. Put another way, this is the smallest ideal in $\underline{A}$ such that $\underline{I_{\mathcal{F}}}(G/H) = A(H)$ whenever $H \in \mathcal{F}$.

We shall see in § 4.3 that this Mackey functor does occur as the homotopy groups of a relatively simple G-spectrum, namely the cofiber of various restriction maps. If H_i runs through all proper subgroups of G, then the quotient Mackey functor has

$$\underline{A/I}(G/H) = \begin{cases} \mathbb{Z} & H = G, \\ 0 & \text{otherwise.} \end{cases}$$

This applies in several cases. For example, if V is a representation such that $V^H \neq \{0\}$ for all proper subgroups H but $V^G = \{0\}$, then $\underline{\pi_0(S^V)}$ is this Mackey functor.

There is another key ideal which shows up, and the associated quotient is also named. Let $\underline{I}(G/H)$ denote the "kernel of the augmentation" $\underline{A}(G/H) \to \underline{A}(G/\{e\}) = \mathbb{Z}$. This map takes a G-set X to its cardinality. Since the map is the restriction, these kernels restrict to each other, and the condition that the transfer is a module map ensures that they are closed under transfers, and therefore form a Mackey functor.

Definition 4.5. The augmentation ideal is the Mackey functor ideal $\underline{I}$ given by the kernel of the augmentation map.

The constant Mackey functor $\underline{\mathbb{Z}}$ is the quotient $\underline{A}/\underline{I}$. The value at G/H is $\mathbb{Z}$ and all restriction maps are isomorphisms.

The Mackey functor $\underline{\mathbb{Z}}$ will play a central role in this paper, as computations with it are especially doable (unlike, say, with a more general Mackey functor). We will soon see how to compute the homology of any representation sphere S^V with coefficients in $\underline{\mathbb{Z}}$.

4.2. $RO(G)$-Graded homotopy groups. One of the advantages of an honest equivariant setting is the presence of a large class of invertible spheres. If W is a virtual representation of G, then we have a sphere S^W, and the assignment $W \mapsto S^W$ is monoidal where in the target we take the smash product. This gives us a large collection of homotopy groups.

Definition 4.6. The $RO(G)$-graded homotopy groups of X, denoted $\pi_\star(X)$, are given by

$$[S^W, X], \quad W \in RO(G).$$

Just as before we can prolong this to a Mackey functor:

$$\underline{\pi_V(X)}(G/H) = [G/H_+ \wedge S^V, X].$$

This approach was pioneered by Lewis who used it to compute the $RO(G)$-graded homology of projective spaces [**21**]. We will use this extensively to determine the homotopy groups of our spectrum $\tilde{\Omega}$.

The representation ring has a distinguished sub semiring: isomorphisms classes of genuine representations. We shall call this the "positive" part of the representation ring. Similarly, any virtual representation of the form $-V$ will be called "negative". Since the positive and the negative parts of the representation ring are sub-semirings, if E is an equivariant ring spectrum, $\pi_{\star\geq 0}E$ and $\pi_{\star\leq 0}E$ are rings graded on honest representations. It is often easiest to give generators and relations for these, rather than $\pi_\star E$.

We can actually refine this approach slightly. The representation ring itself is a Mackey functor, and we can extend the assignment $W \mapsto S^W$ to a kind of Mackey functor in spaces.

Definition 4.7. If W is a virtual representation of $H \subset G$, let

$$\underline{\pi_W(X)}(G/K) = [G/K_+ \wedge (G_+ \wedge HS^W), X]_G.$$

This will show up only obliquely in our proof. When we look at the cellular decomposition of $MU_{(C_{2^n})}$, we will see extra classes exactly of this form. Having the additional homotopy groups would allow a cleaner description of the homology of $MU_{(C_{2^n})}$. Additionally, this extra structure underscores the existence of additional restrictions and transfer maps. These transfer maps (especially related to the sign representations of subgroups of C_{2^n}) play a large role in determining differentials on normed classes.

4.3. Some equivariant homotopy groups. Though the equivariant homotopy groups of spheres are monstrously difficult to compute, there are several cases where we can do so: we can compute $\underline{\pi_0(S^V)}$ for any representation V. We begin with two observations which give the full form. The first is essentially a restatement of tom Deick's computations.

OBSERVATION 4.8. *The abelian group $\pi_0^G S^0$ is the free abelian group generated by $[G/H]$ where H ranges over all subgroups of G. The element $[G/H]$ is the image of 1 under the transfer map*

$$\pi_0^H S^0 = A(H) \to A(G) = \pi_0^G S^0.$$

for more general representations, we look at fixed points.

OBSERVATION 4.9. *If H is a subgroup of G for which $V^H \neq \{0\}$, then*

$$\pi_0^{H'} S^V = 0$$

for any $H' \subset H$.

This is simply because to understand $\underline{\pi_0 S^0}(G/H')$, it suffices to restrict attention to H'-Spectra and compute the Mackey functor π_0 there. Since we have a trivial summand, the computation reduces to that of $\pi_0 S^1$.

The condition $V^H \neq \{0\}$ is clearly closed under subgroups of H and under conjugation. Thus for any representation V, we have a family $\mathcal{F}_V$: the subgroups of G with non-trivial fixed points. The previous observations show the following result.

PROPOSITION 4.10. *As $\underline{A}$-modules,*

$$\underline{\pi_0 S^V} \cong \underline{A/I_{\mathcal{F}_V}}.$$

There is a distinguished generator of $\underline{\pi_0 S^V}$.

DEFINITION 4.11. Let a_V denote the inclusion of $S^0 = \{0, \infty\}$ into S^V.

This class is a kind of Euler class [**13**]. Multiplication by a_V gives us a map $\underline{A} \to \underline{\pi_0 S^V}$, and this realizes the above isomorphism. The following proposition is immediate from either the definition or our results about $\pi_0 S^V$, but it will help us greatly in manipulating classes.

PROPOSITION 4.12.

(1) *If 1 denotes $\mathbb{R}$ with the trivial action, then $a_1 = 0$.*
(2) *For any V and W, $a_{V\oplus W} = a_V \cdot a_W$.*
(3) *If $H \subset G$, then $Res_G^H(a_V) = a_{Res_H V}$.*
(4) *If $H \subset G$, then $N_H^G(a_V) = a_{Ind_H^G V}$.*

In particular, we see that it suffices to consider a_V where V ranges over irreducible representations (though it will be notationally simpler to consider all such classes).

These classes are actually torsion-free in the homotopy groups of spheres. Their image in $\pi_\star H\underline{\mathbb{Z}}$, which we will also denote by a_V is in general not. We can spell this out more explicitly.

PROPOSITION 4.13. *If* $Res^H_G(a_V) = 0$, *then* $|G/H|{\cdot}a_V = 0$ *in* $\pi_0 S^V \wedge H\underline{\mathbb{Z}}$.

PROOF. The Hurewicz map on π_0 is the reduction modulo the augmentation ideal in the Burnside ring. In this ideal we have the obvious element $[G/H] - |G/H|[G/G]$. □

For cyclic 2-groups (and more generally for cyclic p-groups), these classes have a very close connection to geometric fixed points and to the various families of subgroups one can consider.

THEOREM 4.14. *For any* X, *we have*

$$\pi_\star(\tilde{E}\mathcal{F}_m \wedge X) = a^{-1}_{\lambda(p^{m-1})}\pi_\star(X).$$

PROOF. Since $\tilde{E}\mathcal{F}_m$ is the cofiber of the natural map $E\mathcal{F}_{m+} \to S^0$, we can identify it with

$$\lim_{\to} S^{k\lambda(p^{m-1})},$$

where the maps in the colimit are the suspensions of $a_{\lambda(p^{m-1})}$. The result follows. □

COROLLARY 4.15. *The homotopy groups of* $\Phi^G(X)$ *are given by the integer graded part of* $a_\sigma^{-1}\pi_\star(X)$.

As was remarked above, the space $E\mathcal{F}_m$ is actually just a homotopy type. We could use any family of representations that had this property. In particular, we could add any collection of representations with a smaller stabilizer subgroup (named those with a smaller power of λ).

COROLLARY 4.16. *If we invert* $a_{\lambda(p^{m-1})}$ *then we invert* $a_{\lambda(p^k)}$ *for all* $k \leq m-1$.

COROLLARY 4.17. *Inverting* a_σ *inverts* a_V *for any representation* V *with trivial fixed points.*

This will make computations in the slice tower much simpler. In particular, we can use this to greatly reduce the complexity of the computations involving differentials in the slice tower.

5. The slice tower and the slices of $MU_{(C_{2^n})}$

5.1. The homology of $MU_{(C_{2^n})}$. We begin with the following C_2-equivariant generalization of the classical splitting of $MU \wedge MU$ as MU-module spectra.

PROPOSITION 5.1. *As* $MU_{\mathbb{R}}$*-modules, we have a splitting*

$$MU_{\mathbb{R}} \wedge MU_{\mathbb{R}} \simeq MU_{\mathbb{R}} \wedge \bigvee_p S^{\frac{|p|}{2}\rho_2},$$

where p *ranges over all monic monomials in* MU_*.

Since it will show up a very great deal, we give names to particular wedges of representation spheres.

DEFINITION 5.2. Let

$$W = \bigvee_p S^{\frac{|p|}{2}\rho_2},$$

where p ranges over all monic monomials in MU_*.

The previous proposition gives us two immediate corollaries. First, it allows us to conclude the $\underline{RO(G)}$-graded homology of $MU_{\mathbb{R}}$.

COROLLARY 5.3. *We have a splitting*

$$H\underline{\mathbb{Z}} \wedge MU_{\mathbb{R}} \simeq H\underline{\mathbb{Z}} \wedge W.$$

This can actually be shown geometrically using a Schubert cell decomposition of the Grassman manifolds which occur as the spaces in the $MU_{\mathbb{R}}$-spectrum. This splitting underscores also the importance of the regular representation sphere for $MU_{\mathbb{R}}$, and in our construction of the slice tower, we will capitalize on this.

The second corollary allows us to increase the size of the group.

COROLLARY 5.4. *We have a splitting of $MU_{(C_{2^n})}$-module spectra*

$$MU_{(C_{2^n})} \wedge MU_{(C_{2^n})} \simeq MU_{(C_{2^n})} \wedge N_{C_2}^{C_{2^n}} W.$$

Our understanding of the norm of a wedge allows us to reformulate this last part into a more explicit form.

THEOREM 5.5. *We have an equivariant decomposition*

$$N_{C_2}^{C_{2^n}} W = \bigvee_{p \in I} C_{2^n}/H_{p+} \wedge S^{\frac{|p|}{|H_p|}\rho_{H_p}},$$

where I is the orbits of monomials in $\pi_*^{\{e\}} MU_{(C_{2^n})} \otimes \mathbb{Z}/2$ and where H_p is the stabilizer subgroup of p.

REMARK 5.6. Since we are working modulo 2, it is immediate that $C_2 \subset H_p$ for all p. Thus we never have free summands in this wedge.

This in particular gives us a splitting of $H\underline{\mathbb{Z}} \wedge MU_{(C_{2^n})}$, and we see that here the important cells are those of the form

$$C_{2^{n+1}}/H_+ \wedge S^{k\rho_H}$$

for various non-trivial subgroups H. This slice tower capitalizes on this. By judiciously choosing an equivariant filtration, we can ensure that the associated graded of $MU_{(C_{2^n})}$ is exactly $H\underline{\mathbb{Z}} \wedge MU_{(C_{2^n})}$. As we shall see in the next section, this is very computable.

5.2. Slice filtration. A key step in our solution is the production of a new equivariant filtration, the slice filtration. This is a refinement of the non-equivariant Postnikov tower, but it groups the homotopy according the regular representation cells. In fact, there is a choice here of "slice" cells, and different choices produce different slice spectral sequences. The choice we use plays nicely with the spectra $MU_{(C_{2^n})}$, allowing us to identify the slice associated graded and to use it to compute equivariant homotopy groups.

DEFINITION 5.7. Any generalized representation sphere of the form $G_+ \wedge_H S^{k\rho_H - \epsilon}$, where ϵ is 0 or 1, is a slice cell. The dimension of the slice cell $G_+ \wedge_H S^{k\rho_H-\epsilon}$ is $k|H| - \epsilon$.

The inclusion of the factor ϵ seems initially somewhat confusing. This is actually also dictated by the Schubert cell decomposition of $MU_{(C_{2^n})}$. We know that in homology we see cells of the desired form (without any ϵ factor), but this means that they arise as cofibers of maps from a single desuspension. Having these desuspensions allows us to also control these maps, and not just the cells.

Associated to these collections of cells, we can build localizing subcategories.

DEFINITION 5.8. Let $\mathcal{T}_n$ denote the full subcategory of G-spectra built by closing the category of all slice cells of dimension greater than n under colimits and extensions.

In other words, we consider all spectra we can build out of the slice cells of dimension at least n under iterated colimits and under extensions. We do not allow require (nor do we want to allow) arbitrary limits or fibers. This is a big difference between our slice filtration and the motivic one due to Voevodsky: the motivic slice filtration is a filtration of triangulated subcategories whereas ours in not.

DEFINITION 5.9. The nth slice truncation functor P^n is the Dror-nullification functor associated to the subcategory $\mathcal{T}_n$.

This is most easy described via its universal properties. We have P^n and a natural transformation $1 \to P^n$ such that the space of maps $Map(W, P^n(X))$ is equivariantly contractible for any slice cell W of dimension at least n and such that $P^n(X)$ is initial amongst spectra with this property [**12**].

It is easy to check that $G_+ \wedge_H S^{k\rho_H-\epsilon}$ is $(k-\epsilon)$-connected if $k \geq 0$ and $(k|H|)$-connected if $k < 0$. This give us a coarse lower bound on the connectivity of any element in $\mathcal{T}_n$: $n/|G| - 1$. A general property of nullification then ensures the following fact (which for easy, we state only for $n > 0$).

PROPOSITION 5.10. *The map* $X \to P^n(X)$ *is* $(n/|G| - 1)$*-connected.*

Dually, it is easy to show that for all H, $G_+ \wedge_H S^n \in \mathcal{T}_n$. These are exactly the cells which are killed to build the equivariant Postnikov tower, we conclude the following.

PROPOSITION 5.11. *For all n and for all $k > n$,*

$$\underline{\pi_k}(P^n(X)) = 0.$$

DEFINITION 5.12. The slice tower of X is the iterated nullification tower

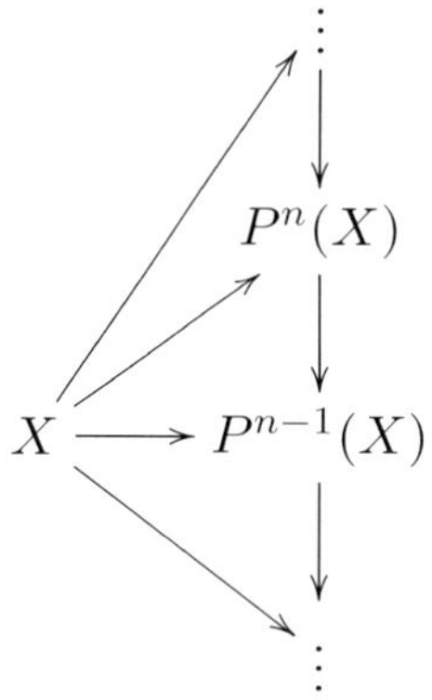

The slice spectral sequence is the spectral sequence associated to this filtration.

Proposition 5.10 ensures that the map from X to the limit of the slice tower is a weak equivalence. Proposition 5.11 ensures that the colimit of the slice tower is contractible. Thus the slice spectral sequence has good convergence properties.

The E_1-term of this spectral sequence is given by the Mackey functor homotopy groups of the fibers $P^n(X) \to P^{n-1}(X)$. These fibers are the slices of X.

DEFINITION 5.13. The n-slice of X is the homotopy fiber $P^n(X) \to P^{n-1}(X)$.

In general, determining the slices of X is regrettably difficult. There are two cases where we can do so.

PROPOSITION 5.14. *If T is a slice cell, then $T \wedge H\underline{\mathbb{Z}}$ is a $\dim(T)$-slice.*

This is an essentially immediately computation of equivariant homology groups, and we postpone it until a later section.

COROLLARY 5.15. *If again*

$$W = \bigvee_p S^{\frac{|p|}{2}\rho_2},$$

then

$$W \wedge H\underline{\mathbb{Z}} \quad \textit{and} \quad \left(N_{C_2}^{C_{2^n}} W\right) \wedge H\underline{\mathbb{Z}}$$

are wedges of slices.

PROPOSITION 5.16. *The zero slice of S^0 is $H\underline{\mathbb{Z}}$.*

PROOF. We prove this for cyclic p-groups, leaving the full case to the larger paper. The proof is by induction on the order of the group. We know that the slice tower is a refinement of the Postnikov tower. In particular, for the zero-slice, this means two things. First, the zero-slice of S^0 is of the form $H\underline{M}$ for some $\underline{M}$ and second, the underlying spectrum must be $H\mathbb{Z}$. In particular, though transfers of the element $1 \in \underline{M}(G/\{e\})$ might be killed, nothing that restricts to it can be.

We will show by induction on the group that any element in the augmentation ideal $\underline{I}(G/G)$ can be realized by a map from some wedge of slice cells. The case for G/H is handled identically. It suffices, by induction for transfer reasons, to then only consider elements in the augmentation ideal that are not in the image of the transfer from any subgroup. For a general group, these are generated by classes $[G/H]-|G/H|$ for H a maximal proper subgroup of G. In the case of a cyclic group of prime power order, these are especially simple: there is only one such H that is also a cyclic group of prime power order.

Since the restriction of $\alpha = [C_{p^n}/C_{p^{n-1}}] - p$ to $C_{p^{n-1}}$ is zero, we know that the composition

$$C_{p^n+} \wedge_{C_{p^{n-1}}} S^0 \to S^0 \xrightarrow{\alpha} H\underline{A}$$

is null-homotopic. Our understanding of how to build the a cell structure for $S^{\rho-1}$ shows that the 1-cells are attached by exactly the map

$$C_{p^n+} \wedge_{C_{p^{n-1}}} S^0 \to S^0,$$

and so α extends over the 1-skeleton. Since $H\underline{A}$ is an Eilenberg-Mac Lane spectrum, we conclude that α extends over $S^{\rho-1}$. □

A similar proof actually shows that the category of zero slices is the category of Mackey functors for which all restriction maps are monomorphisms. This is true for an arbitrary finite group.

PROPOSITION 5.17. *Let V be a representation, and let $\mathcal{F}_V$ be the family of subgroups H of G such that $V^H \neq 0$. Then for any Mackey functor $\underline{M}$ we have*

$$[S^V, H\underline{M}] = \{x \in \underline{M}(G/G) | Res_G^H(x) = 0, \forall H \in \mathcal{F}_V\}.$$

PROOF. Since $H\underline{M}$ is an Eilenberg-Mac Lane spectrum, we know

$$[S^V, H\underline{M}] = [H\underline{\pi_0 S^V}, H\underline{M}] = Hom_{\underline{A}}(\underline{\pi_0 S^V}, \underline{M}).$$

The result now follows from our earlier analysis of $\pi_0 S^V$. □

For our purposes, the most important example where we can determine the slice associated graded is $MU_{(C_{2^n})}$.

THEOREM 5.18 (the Slice Theorem). *There is an equivariant filtration of $MU_{(C_{2^n})}$ (the* slice *filtration) such that the associated graded is*

$$H\underline{\mathbb{Z}} \wedge \bigvee_{p \in I} C_{2^n}/H_{p+} \wedge S^{\frac{|i|}{|H_p|}\rho_{H_p}},$$

where I is the orbits of monomials in $\pi_*^{\{e\}} MU_{(C_{2^n})} \otimes \mathbb{Z}/2$ and where H_p is the stabilizer subgroup of p.

Rather than give a complete proof, we will sketch the argument. We begin with a statement about the underlying homotopy of $MU_{(C_{2^n})}$.

PROPOSITION 5.19. *We can find generators of $\pi_*^{\{e\}} MU_{(C_{2^n})}$ such that as an equivariant algebra*

$$\pi_*^{\{e\}} MU_{(C_{2^n})} = \mathbb{Z}[r_1, \ldots, \gamma^{2^{n-1}-1} r_1, r_2, \ldots],$$

where γ, the generator of C_{2^n}, acts in the obvious way, and where $\gamma^{2^{n-1}} r_i = (-1)^i r_i$. The degree of r_i is $2i$.

For $n = 1$, these can be chosen to be the usual classes. For $n = 2$, we can choose these to (up to decomposables) be the usual classes t_i.

These underlying homotopy classes are actually the restriction of C_2-equivariant maps from non-trivial representation spheres, and from this we get a good description of equivariant generators in general.

PROPOSITION 5.20. *There are C_2-equivariant maps*

$$\bar{r}_i \colon S^{i\rho_2} \to MU_{(C_{2^n})}$$

such that the underlying map is r_i.

When we need to remember the ambient group, we will further adorn $\bar{r}_i$ with a second subscript: $\bar{r}_{i,n}$ is $\bar{r}_i$ for C_{2^n} and $MU_{(C_{2^n})}$. The left unit allows us to find $\bar{r}_{i,k}$ in $\pi_{i\rho_2} MU_{(C_{2^n})}$ for all $k \leq n$.

Applying the C_{2^n}-action then gives us classes $\gamma^j \bar{r}_i$.

DEFINITION 5.21. Let $S[\bar{r}_i]$ denote the free A_∞-ring spectrum on $S^{i\rho_2}$.

Thus

$$S[\bar{r}_i] = \bigvee_{j \geq 0} S^{ij\rho_2},$$

and a map of associative algebras $S[\bar{r}_i] \to X$ is given by an element $\pi_{i\rho_2} X$. As the name implies, there is for each i a distinguished map $S[\bar{r}_i] \to MU_{(C_{2^n})}$. Since $MU_{(C_{2^n})}$ is a commutative ring spectrum, we can smash all of these algebras together and get a map of associative algebras

$$\bigwedge_{k \geq 1} S[\bar{r}_k] \to MU_{(C_{2^n})}.$$

The underlying map for this exactly picks out the subalgebra generated by the generators r_k in homotopy.

If we apply the norm, followed by the canonical map $N_{C_2}^{C_{2^n}} MU_{(C_{2^n})} \to MU_{(C_{2^n})}$, then we get a map of associative algebras

$$N(\bigwedge_{k\geq 1} S[\bar{r}_k]) \to MU_{(C_{2^n})}.$$

Let R denote the associative algebra $N(\bigwedge_{k\geq 1} S[\bar{r}_k])$.

Since $S[\bar{r}_k]$ is a wedge of equivariant spheres, we can easily understand

$$R = \bigwedge_{k\geq 1} N(S[\bar{r}_k]).$$

PROPOSITION 5.22. *The homology of the underlying spectrum for R is*

$$\mathbb{Z}[r_1, \dots, \gamma^{2^{n-1}-1} r_1, r_2, \dots],$$

where r_i is the fundamental class of the associated sphere, and where $|r_i| = 2i$, and the underlying map on homotopy

$$\pi_*^{\{e\}} R \to \pi_*^{\{e\}} MU_{(C_{2^n})}$$

is surjective, sending fundamental classes to the identically named elements.

In the group action, $\gamma^{2^{n-1}}$ acts as $(-1)^i$ on r_i. In fact, this sign action is recorded in the equivariant homotopy by the fact that the conjugation map $S^{\rho_2} \to S^{\rho_2}$ had degree -1. We also know that every class in the norm of our wedge has a stabilizer group containing C_2. Putting these two facts together leads us to consider orbits of monomials in the underlying homotopy of $MU_{(C_{2^n})}$. Associated to each orbit, we can attach the stabilizer subgroup modulo 2, stripping out the pesky sign problem. For easy of notation, if p is a monomial, let H_p denote this stabilizer subgroup.

This gives us exactly the equivariant homotopy type of R: it is a wedge of induced spheres of the form

$$C_{2^n +} \wedge_{H_p} S^{\frac{|p|}{|H_p|}\rho_{H_p}},$$

where p ranges over a set of representatives for the orbits of monomials in the underlying homotopy of $MU_{(C_{2^n})}$.

The reader will at this point no doubt recognize that the claim of the necessary theorem is that $MU_{(C_{2^n})}$ has an equivariant filtration for which the associated graded is $H\underline{\mathbb{Z}} \wedge R$. We can achieve this by first recognizing that since R is essentially a polynomial algebra in spectra (though we stress that the variables do not commute with themselves in the sense that there are no power operations beyond the pth power), we have a filtration by degree, with the "degree at least n" bimodule being denoted M_n. Using the map of associative algebras $R \to MU_{(C_{2^n})}$, we get a decreasing filtration of

$MU_{(C_{2^n})} = MU_{(C_{2^n})} \wedge_R R$ by considering $MU_{(C_{2^n})} \wedge_R M_n$. The associated graded for this filtration is simply

$$\left(MU_{(C_{2^n})} \wedge_R S^0\right) \wedge R.$$

This gives rise to by far the most difficult theorem in our proof: the Reduction Theorem. Since the proof is somewhat technical, we will not reproduce it here.

THEOREM 5.23 (the Reduction Theorem). *We have an equivariant equivalence*

$$MU_{(C_{2^n})} \wedge_R S^0 \simeq H\underline{\mathbb{Z}}.$$

In fact, the map $MU_{(C_{2^n})} \wedge_R S^0 \to H\underline{\mathbb{Z}}$ realizing this equivalence is the 0th Postnikov section of $MU_{(C_{2^n})}$. We should interpret the Reduction Theorem as an equivariant analogue of the fact that MU modulo all of its generators in non-zero degree is $H\mathbb{Z}$. As a consequence of the Reduction Theorem, we also see that $MU_{(C_{2^n})} \wedge_R S^0$ inherits an algebra structure. This is not immediate, since R is not commutative.

The astute reader will notice that at no point in the argument up to this point have we used the slice filtration. It is precisely in our proof of the reduction theorem that having the full slice story becomes most useful.

With the Reduction Theorem, the description of the associated graded of $MU_{(C_{2^n})}$ is complete. Before focusing on the specific computations, we give a brief corollary.

COROLLARY 5.24. *For any k, there is an equivariant filtration of $\Sigma^{k\rho_{2^n}} MU_{(C_{2^n})}$ such that the associated graded is*

$$H\underline{\mathbb{Z}} \wedge \bigvee_{p \in I} C_{2^n}/Hp+ \wedge S^{\frac{|i|+k2^n}{|H_p|}\rho_{H_p}},$$

where again I is the orbits of monomials in $\pi_*^{\{e\}} MU_{(C_{2^n})} \otimes \mathbb{Z}/2$ and where H_p is the stabilizer subgroup of p.

This is immediate from smashing the filtration of $MU_{(C_{2^n})}$ with $S^{k\rho_{2^n}}$. Here we have used that the restriction of the regular representation of C_{2^n} to any subgroup is a sum of copies of the regular representation. In particular, we see that the form of the summands,

$$H\underline{\mathbb{Z}} \wedge (C_{2^n+} \wedge_{C_{2^k}} S^{j\rho_{2^k}}),$$

is unchanged!

6. The homology of a point, and the Gap theorem

For cyclic groups of prime-power order and for a virtual representation spheres S^V, it is easy to compute the Mackey functor homotopy group $\underline{\pi_*(S^V \wedge H\underline{\mathbb{Z}})}$. Using the linear ordering of the subgroups, we can build a triangularization of S^V which gives us an especially simple chain complex.

For ease of exposition, we will restrict attention to cyclic 2-groups. Everything said here applies equally well to cyclic p-groups. Our exposition is greatly streamlined by the following obvious fact: the subgroups of a cyclic p-group are linearly ordered.

The analysis begins with a decomposition of S^V into cells for V an honest representation. For this, we can actually understand this *in spaces.* This gives us, by equivariant Spanier-Whitehead duality, a cellular decomposition of S^{-V}, and then by smashing, we get one for any virtual representation. This translates into our chain complexes by simply requiring us to form the associated cochain complex for S^{-V} and then to tensor things together for a more general case. We remark that for honest representations, this chain complex is much smaller than the one you get by decomposing V into irreducible representations and tensoring the associated complexes together.

For the rest of this section, let V be a fixed representation of C_{2^n}.

6.1. The Bredon homology of S^V. We can decompose V into a direct sum of representations with a fixed stabilizer subgroup:

$$V = \bigoplus_{k=0}^{n} V_k,$$

where V_k is a sum of representations of the form $\lambda(j)$ with the 2-adic valuation of j equal to $n-k$. Thus V_i is stabilized by $C_{2^{n-i}}$, and conversely, the fixed points of $C_{2^{n-i}}$ is

$$V^{C_{2^{n-i}}} = \bigoplus_{k=0}^{i} V_k.$$

This actually immediately lets us built a cellular decomposition. Since any representation spheres $S^{\lambda(i)}$ and $S^{\lambda(i')}$ are equivalent if i and i' have the same 2-adic valuation, we can simple consider

$$V = m_0 \oplus m_1\sigma \oplus m_2\lambda(2^{n-2}) \oplus \cdots \oplus m_n\lambda.$$

This particular ordered decomposition of V into irreducible representations with shrinking stabilizer groups gives our cells. For ease on the reader's eyes, let

$$M_k = \dim V^{C_{2^{n-k}}} = m_0 + m_1 + 2m_2 + \cdots + 2m_k.$$

PROPOSITION 6.1. *There is a cell decomposition of S^V of the form*

$$S^{m_0} \cup \bigcup_{j_1=1}^{m_1} C_{2^n+} \wedge_{C_{2^{n-1}}} e^{M_0+j_1} \cup \cdots \cup \bigcup_{j_n=1}^{2m_n} C_{2^n+} \wedge e^{M_{n-1}+j_n}.$$

Moreover, the subcomplexes

$$S^{m_0} \cup \bigcup_{j_1=1}^{k} C_{2^n+} \wedge_{C_{2^{n-1}}} e^{m_0+j_1}$$

and

$$S^{m_0} \cup \bigcup_{j_1=1}^{m_1} C_{2^n +} \wedge_{C_{2^{n-1}}} e^{m_0+j_1} \cup \cdots \cup \bigcup_{j_k=1}^{2\ell} C_{2^n +} \wedge_{C_{2^{n-k}}} e^{M_{k-1}+j_k}.$$

are the representation spheres

$$S^{m_0+k\sigma} \text{ and } S^{m_0+m_1\sigma+\cdots+\ell\lambda(2^{n-k})}$$

respectively.

The second half essentially tells us how to build these. We attach cells with increasingly smaller stabilizer groups to the representation spheres we have already built. We can therefore understand the cell structure by understanding how to build $S^{V\oplus\lambda(2^k)}$ with S^V as the co-dimension 2-skeleton for representations V with stabilizer containing C_{2^k}.

This is easiest to visualize with a picture. In Figure 3, the z-axis represents the representation V, while the (x, y)-plane (herein depicted as $\lambda(2^{n-2})$) is $\lambda(2^k)$. The white dot in the center is the origin, and the shading on the semi-planes indicates the contribution of the group action on the vertical component (in this case, flipping back and forth).

We have a natural cell structure for $S^{\lambda(2^k)}$:

$$S^0 \cup C_{2^n +} \wedge_{C_{2^{n-k}}} e^1 \cup C_{2^n +} \wedge_{C_{2^{n-k}}} e^2.$$

We can easily visualize this cell structure on the Riemann sphere. The 0-cells are the origin and the point at infinity, the 1-cells are the great semi-circles through the zero cells and the $2^{k\text{th}}$-roots of unity, and the 2-cells interpolate between these. This is depicted in Figure 4 for $k = 3$.

To build our cell structure on $S^{V\oplus\lambda(2^k)}$, we smash this cell structure with the spheres already built. Since the stabilizer subgroup of V contains $C_{2^{n-k}}$, we know that

$$S^V \wedge \left(C_{2^n +} \wedge_{C_{2^{n-k}}} e^1\right) = C_{2^n +} \wedge_{C_{2^{n-k}}} S^V \wedge e^1 = C_{2^n +} \wedge_{C_{2^{n-k}}} e^{1+\dim V},$$

Thus we can ignore the codimension 1-skeleton of S^V after smashing with $C_{2^n}/C_{2^{n-k}}$, and we produce the desired cell structure on $S^{V\oplus\lambda(2^k)}$.

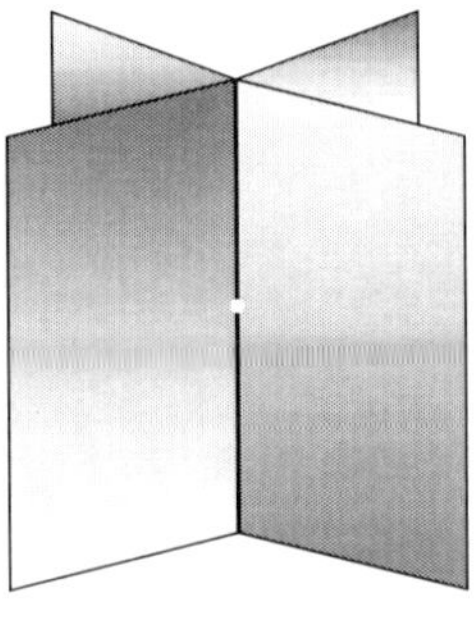

FIGURE 3. Adding New Cells

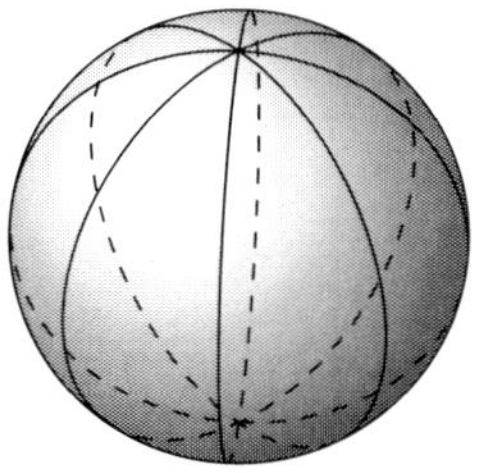

FIGURE 4

This gives us very simple chain complexes to compute the Bredon homology.

DEFINITION 6.2. If $V = m_0 + m_1\sigma + m_2\lambda(2^{n-2}) + \cdots + m_n\lambda$, then let $C(V)$ denote the chain complex

$$C(V)_j = \begin{cases} \mathbb{Z}[C_{2^n}/C_{2^{n-k}}] & M_{k-1} < j \leq M_k \\ 0 & \text{otherwise,} \end{cases}$$

where the differentials are those equivariant maps such that the restriction of the complex to those cells of dimension at most $M_{k-1}+2j$ is quasi-isomorphic to the chains on $S^{M_{k-1}+2j}$.

The complex $C(V)$ is the equivariant cellular chain complex. If we take fixed points for various subgroups and take homology, then we recover the Bredon homology groups, by definition.

PROPOSITION 6.3. *As Mackey functors,*

$$\underline{H_n(S^V)}(G/H) := \underline{\pi_n(S^V \wedge H\underline{\mathbb{Z}})}(G/H) = H_n(C(V)^H).$$

We can rewrite the final step as saying that homotopy classes of equivariant maps from the chain complex that is $\mathbb{Z}[G/H]$ in dimension n to $C(V)$ is the same as the homology of the fixed points of the complex.

The most important cases are regular representations spheres, and we spell out several cases for the reader's convenience. From these, the general pattern easily follows.

EXAMPLE 6.4. For C_4, we have

$$C(\rho_4) = \mathbb{Z}[C_4/C_4] \xleftarrow{\nabla} \mathbb{Z}[C_4/C_2] \xleftarrow{1-\gamma} \mathbb{Z}[C_4] \xleftarrow{1+\gamma} \mathbb{Z}[C_4]\ ,$$

where ∇ is the "fold" map $\nabla\colon \mathbb{Z}^2 \to \mathbb{Z}$. We also have

$$C(2\rho_4) = \mathbb{Z} \xleftarrow{\nabla} \mathbb{Z}^2 \xleftarrow{1-\gamma} \mathbb{Z}^2 \xleftarrow{1+\gamma} \mathbb{Z}^4 \xleftarrow{1-\gamma} \mathbb{Z}^4 \xleftarrow{Tr} \mathbb{Z}^4 \xleftarrow{1-\gamma} \mathbb{Z}^4\ ,$$

where we have abreviated $\mathbb{Z}[C_4/H]$ to $\mathbb{Z}^{(4/|H|)}$ and where $Tr = (1 + \gamma + \gamma^2 + \gamma^3)$.

There is an essential fact that underlies all of what follows. If V contains σ, then $C(V)$ always begins

$$\mathbb{Z}[C_{2^n}/C_{2^n}] \xleftarrow{\nabla} \mathbb{Z}[C_{2^n}/C_{2^{n-1}}].$$

This quite simple result is the basis for the gap theorem!

6.2. The Bredon homology of S^{-V}. Since S^{-V} is the equivariant Spanier-Whitehead dual of S^V, the Bredon chains are the dual to the chains on S^V:

$$C(-V) := Hom(C(V), \mathbb{Z}),$$

where $\mathbb{Z}$ has the trivial action. The two most important examples are $C(\rho_{2^n})$ and $C(2\rho_{2^n})$. We again illustrate these for C_4.

EXAMPLE 6.5. We have

$$C(-\rho_4) = \mathbb{Z}[C_4/C_4] \to \mathbb{Z}[C_4/C_2] \to \mathbb{Z}[C_4] \to \mathbb{Z}[C_4],$$

where now the first map is the diagonal $\mathbb{Z} \to \mathbb{Z}^2$, and the remaining ones are as in the positive case.

We also have

$$C(-2\rho_4) =$$
$$\mathbb{Z}[C_4/C_4] \to \mathbb{Z}[C_4/C_2] \to \mathbb{Z}[C_4/C_2] \to \mathbb{Z}[C_4] \to \mathbb{Z}[C_4] \to \mathbb{Z}[C_4] \to \mathbb{Z}[C_4],$$

and again the only change is that the first map is the diagonal map.

PROPOSITION 6.6. *If V contains a copy of σ, then the two non-zero terms of largest degree in $C(-V)$ are*

$$\mathbb{Z}[C_{2^n}/C_{2^n}] \xrightarrow{\Delta} \mathbb{Z}[C_{2^n}/C_{2^{n-1}}],$$

where Δ is the diagonal map $\mathbb{Z} \to \mathbb{Z}^2$.

6.3. The Gap theorem. We begin with a few simple homotopy groups.

LEMMA 6.7. *For any $k > 0$ and any j, we have*

$$\pi_{-3,-2,-1}\big(H\underline{\mathbb{Z}} \wedge C_{2^n+} \wedge_{C_{2^k}} S^{j\rho_{2^k}}\big) = 0.$$

PROOF. By the natural adjunctions between induction and restriction, it will suffice to show that for any j

$$\pi_{-3,-2,-1} H\underline{\mathbb{Z}} \wedge S^{j\rho_{2^n}} = 0.$$

If $j \geq 0$, then elementary connectivity arguments show this is so. For $j < -3$, the $(-j)$-connectivity of $S^{\ j\rho_{2^n}}$ produces the desired co-connectivity. For j between -3 and -1 inclusive, Proposition 6.6 shows that in dimensions j and $j-1$, $C(j\rho_{2^n})$ looks like

$$\mathbb{Z}[C_{2^n}/C_{2^n}] \xrightarrow{\Delta} \mathbb{Z}[C_{2^n}/C_{2^{n-1}}].$$

Upon passage to C_{2^n} fixed points, this complex becomes

$$\mathbb{Z} \xrightarrow{1} \mathbb{Z},$$

and we conclude that in degrees j and $j-1$, the homology is 0. The only remaining case is $j=-1$ and degree -3. It is easy to check that in this case, the map from degree -3 to -4 is injective, resulting in a trivial homology group in degree -3. □

As an immediate consequence of this and the Slice Theorem, we learn a quite surprising fact about the homotopy of the fixed points of $MU_{(C_{2^n})}$.

THEOREM 6.8. *For any k,*

$$\pi_{-3,-2,-1}\Sigma^{k\rho_{2^n}} MU_{(C_{2^n})} = 0.$$

PROOF. For $k \geq 0$, this is immediate, since $MU_{(C_{2^n})}$ is (-1)-connected, and $S^{k\rho_{2^n}}$ is k-connected. For $k<0$, we use Corollary 5.24. Associated to this filtration, we have a spectral sequence computing equivariant homotopy. The E_1-term is the equivariant homotopy of terms of the form

$$H\underline{\mathbb{Z}} \wedge C_{2^n+} \wedge_{C_{2^m}} S^{j\rho_{2^m}},$$

where $m>0$. By Lemma 6.7, $\pi_{-3,-2,-1}$ of this is zero. Since π_{-2} of every term in the associated graded is zero, we conclude the same is true for $\Sigma^{k\rho_{2^n}} MU_{(C_{2^n})}$. □

THEOREM 6.9 (the Gap Theorem). *If $\bar{\Delta}\colon S^{k\rho_{2^n}} \to MU_{(C_{2^n})}$ is any equivariant homotopy class, then*

$$\pi_{-3,-2,-1}\bar{\Delta}^{-1} MU_{(C_{2^n})} = 0.$$

PROOF. Inverting $\bar{\Delta}$ involves forming a directed colimit

$$\cdots \to MU_{(C_{2^n})} \xrightarrow{\bar{\Delta}} \Sigma^{-k\rho_{2^n}} MU_{(C_{2^n})} \xrightarrow{\bar{\Delta}} \cdots.$$

Spheres are still compact objects, and so we conclude

$$\pi_{-3,-2,-1}\bar{\Delta}^{-1} MU_{(C_{2^n})} = \varinjlim \pi_{-3,-2,-1}\Sigma^{mk\rho_{2^n}} MU_{(C_{2^n})} = \varinjlim 0 = 0. \quad \square$$

Thus in a huge family of equivariantly periodic spectra (these with period $k\rho_{2^n}$), we always have a gap in the homotopy of the fixed points. This is actually quite shocking and strong. With very little work, we have produced one of the key steps in our theorem: the Kervaire classes are eventually detected in the zero group.

REMARK 6.10. There are two known antecedents to this result. Atiyah showed that the fixed points of $K_{\mathbb{R}}$ are KO, and the standard computation of the homotopy of KO has $\pi_{-3,-2,-1}KO = 0$. Similarly, unpublished work of Mahowald-Rezk and Goerss-Henn show that in the homotopy of $EO_2(C_4) = E_2^{C_4}$, we have $\pi_{-3,-2,-1}EO_2(C_4) = 0$. Our result shows that these vanishing results are actually quite generic.

7. Some cohomology operations and slice differentials

The Periodicity and Homotopy Fixed Points Theorems rely on a simple cohomological computation. We construct several elements in the equivariant Steenrod algebra in $BP^{(n)}$-modules and we show that these act non-trivially on the powers of $u_{2\sigma}$. Tying these to $N(\bar{r}_i)$ for various i produces for us, by definition, slice differentials.

7.1. Naming elements in Bredon homology. One of the most useful features of this approach is that we can easily understand the full ring structure of $\pi_\star H\underline{\mathbb{Z}}$ (at least the positive part). In particular, we can explicitly name every element, and this allows us to better understand how elements in the slice spectral sequence interact. We saw that the earlier computations that if V is an oriented representation (that is, the generator acts via a degree 1 map on the underlying sphere), then we have a distinguished map $S^{\dim(V)} \to S^V \wedge H\underline{\mathbb{Z}}$.

Definition 7.1. Let u_V denote the orientation map $S^{\dim(V)} \to S^V \wedge H\underline{\mathbb{Z}}$.

Just as with a_V, these satisfy a number of very useful properties.

Proposition 7.2.

(1) *If* 1 *is the trivial representation, then* $u_1 = 1$.
(2) *If* V *and* W *are representation, then* $u_{V\oplus W} = u_V \cdot u_W$.
(3) *If* $H \subset G$, *then* $N_H^G(u_V) \cdot u_{\dim(V) Ind_H^G 1} = u_{Ind_H^G V}$.

The only non-obvious result is the third. This follows immediately from writing out the norm of the source.

This lets us name every element in our chain complexes. We give the generic example for C_8 and various copies of the regular representation thereof.

$$\begin{array}{ccccccc}
8k & & 4k & & 2k & & k \\
\mathbb{Z}^8 \longrightarrow \cdots \longrightarrow \mathbb{Z}^8 & \longrightarrow & \mathbb{Z}^4 \longrightarrow \cdots \longrightarrow \mathbb{Z}^4 & \longrightarrow & \mathbb{Z}^2 \longrightarrow \cdots \longrightarrow \mathbb{Z}^2 & \longrightarrow & \mathbb{Z} \\
| & & | & & | & & | \\
u_{k(\lambda+\lambda(3))} u_{k\lambda(2)} u_{k\sigma} & & a_{k(\lambda+\lambda(3))} u_{k\lambda(2)} u_{k\sigma} & & a_{k(\lambda+\lambda(2)+\lambda(3))} u_{k\sigma} & & a_{k\bar{\rho}_8}
\end{array}$$

This method of naming actually also follows quite quickly from our choice of cell-structure. We restrict attention to orientable V for now and will indicate the simple change for non-orientable V at the end. Since the codimension 2-skeleton of our cell-structure is another representation sphere (and one for which the stabilizer subgroup is larger), it will suffice to understand how to name the elements of $S^{V+\lambda}$ relative to those of S^V. This follows from the following.

Observation 7.3. *The inclusion* $S^V \to S^{V+\lambda}$ *is* a_λ.

Thus for $* \leq \dim(V)$, multiplication by a_λ gives an isomorphism

$$\underline{\pi_* H\underline{\mathbb{Z}} \wedge S^V}(G/H) \to \underline{\pi_* H\underline{\mathbb{Z}} \wedge S^{V+\lambda}}(G/H).$$

It is also immediate from the chain complexes that the $\dim(V)+1$-dimensional homology vanishes. Finally, by definition, the $\dim(V+\lambda)$-homology is generated by $u_{V+\lambda}$. Downward induction on the number of summands then names generators for all homology groups.

So what happens in the non-orientable case? The names here change very slightly. If V is non-orientable, then V contains an odd number of copies of σ, and $V-\sigma$ is orientable. We name the elements of the homology of $S^{V-\sigma}$ in the usual way, and composing with a_σ produces the generators of the homology of S^V. In particular, we see that if V is not orientable, then all homology groups of S^V are 2-torsion.

Though not directly needed, we also list some multiplicative relations in the Bredon homology of a point. These will result is drastic simplifications upon inverting a_σ, which in turn will make our computation of several cohomology operations tractable.

PROPOSITION 7.4. *The elements $u_{\lambda(2^k)}$ are subject to the following relations*

$$2^j a_{\lambda(2^i)} u_{\lambda(2^{i+j})} = a_{\lambda(2^{i+j})} u_{\lambda(2^i)},$$

where here we have used that $\lambda(2^{n-1}) = 2\sigma$.

It is then immediate that for all $k < n-1$,

$$a_\sigma^3 u_{\lambda(2^k)} = 0,$$

since $a_\sigma^2 u_{\lambda(2^k)}$ is divisible by 2.

Though we have not described the other equivariant homology groups, it is not difficult to check that a_σ also acts nilpotently on almost all classes there. More specifically, we have the following.

PROPOSITION 7.5. *As rings,*

$$a_\sigma^{-1}\pi_\star H\underline{\mathbb{Z}} = \mathbb{Z}/2[a_\sigma^{\pm 1}, a_{\lambda(2^{n-2})}^{\pm 1}, \ldots, a_\lambda^{\pm 1}, u_{2\sigma}],$$

and as spectra, $\Phi^G H\underline{\mathbb{Z}} = H\mathbb{Z}[b]$, where $b = u_{2\sigma}/a_{2\sigma}$ is in degree 2.

7.2. Relative cohomology operations. For sake of argument, we assume that $BP^{(n)}$ is an E_∞ ring spectrum. If this is not the case, then we replace it with $MU_{(C_{2^n})}$. While in general there are many more classes that will appear for $MU_{(C_{2^n})}$, we will restrict attention to geometric fixed points (for which things are better behaved). We begin by analyzing the spectrum $H\underline{\mathbb{Z}} \wedge_{BP^{(n)}} H\underline{\mathbb{Z}}$. This is an equivariant E_∞ ring spectrum whose homotopy is the ring of cooperations for homology in the category of $BP^{(n)}$-modules.

In general, the homotopy of this will be very difficult to compute. However, we can identify particular elements which we can easily check to be non-zero by mapping to the geometric fixed points. To begin, we quickly compute the geometric fixed points of this spectrum.

PROPOSITION 7.6. *As commutative ring spectra,*

$$\Phi^G(H\underline{\mathbb{Z}} \wedge_{BP^{(n)}} H\underline{\mathbb{Z}}) = H\mathbb{F}_2[b'] \wedge_{H\mathbb{F}_2} H\mathbb{F}_2[b] = H\mathbb{F}_2[b, b'],$$

where as before $b = b' = u_{2\sigma}/a_{2\sigma}$ *is an element in degree* 2.

Before continuing, we make a very brief digression into names of elements. Just as classically, the classes $\bar{r}_{i,n}$ do not occur in $BP^{(n)}$ unless i is of the form $2^j - 1$. We therefore give new $BP^{(n)}$-names to simplify the writing. Similarly, since in this section n is held fixed, we suppress it from the notation.

DEFINITION 7.7. Let $\bar{v}_i$ denote the element $\bar{r}_{2^i-1,n}$.

Even using $MU_{(C_{2^n})}$, these are the elements that play a key role (since we always have a splitting in equivariant A_∞-algebras that splits off $BP^{(n)}$).

We define our homology classes inductively using the cofiber sequences

$$(G \cdot \bar{v}_i)BP^{(n)} \to BP^{(n)} \to BP^{(n)}/(G \cdot \bar{v}_i).$$

A key observation that makes many of the computations here work is that from the perspective of geometric fixed points, there is no difference between $(G\cdot\bar{v}_i)BP^{(n)}$ and $N(\bar{v}_i)BP^{(n)} = \Sigma^{i\rho_G}BP^{(n)}$, and the map to $BP^{(n)}$ is multiplication by $N(\bar{v}_i)$. This is because all other elements are carried by induced spheres and are therefore wiped out by the geometric fixed points construction.

If we smash this over $BP^{(n)}$ with $H\underline{\mathbb{Z}}$, then we again get a cofiber sequence:

$$(G \cdot \bar{v}_i)BP^{(n)} \wedge_{BP^{(n)}} H\underline{\mathbb{Z}} \to BP^{(n)} \wedge_{BP^{(n)}} H\underline{\mathbb{Z}} \to BP^{(n)}/(G \cdot \bar{v}_i) \wedge_{BP^{(n)}} H\underline{\mathbb{Z}},$$

and from this we can identify various homotopy elements. Let $N(\bar{v}_i)$ denote the homotopy class $S^{(2^i-1)\rho_G} \to (G \cdot \bar{v}_i)BP^{(n)} \wedge_{BP^{(n)}} H\underline{\mathbb{Z}}$ induced by the Thom reduction map $BP^{(n)} \to H\underline{\mathbb{Z}}$ (which is the unit map in the category of $BP^{(n)}$-algebras). A simple check on geometric fixed points shows that this map is not null. However, when we compose with the map to $BP^{(n)} \wedge_{BP^{(n)}} H\underline{\mathbb{Z}}$, then this map becomes null.

DEFINITION 7.8. Let $\tau_i \colon S^{(2^i-1)\rho_G+1} \to BP^{(n)}/(G\cdot\bar{v}_i)\wedge_{BP^{(n)}} H\underline{\mathbb{Z}}$ denote a preimage of $N(\bar{v}_i)$ in the long exact sequence in homotopy induced by the cofiber sequence.

Let $\tau_i \colon S^{(2^i-1)\rho_G+1} \to H\underline{\mathbb{Z}} \wedge_{BP^{(n)}} H\underline{\mathbb{Z}}$ also denote the image of τ_i into the colimit.

Obviously there is substantial ambiguity in this definition. However, for our purposes we need most understand what happens modulo the elements annihilated by a_σ. For this, the following proposition shows there is not only is there no ambiguity but also the classes are non-zero.

PROPOSITION 7.9. *The image of* $a_{(2^i-1)\bar{\rho}_G}\tau_i$ *in* $\pi_*\Phi^G(H\underline{\mathbb{Z}} \wedge_{BP^{(n)}} H\underline{\mathbb{Z}})$ *is* b'^{2^i}.

PROOF. This is immediate from applying geometric fixed points to the defining cofiber sequences. □

In fact, this proposition is essentially equivalent to the Reduction theorem. The key step is that we can get $H\underline{\mathbb{Z}}$ from $BP^{(n)}$ by killing classes, and the argument relies on a computation. This computation is at its heart equivalent to knowing that the classes τ_i map to non-trivial classes in $H\underline{\mathbb{Z}} \wedge_{BP^{(n)}} H\underline{\mathbb{Z}}$.

Since this is a map of rings, we note that there are non-trivial multiplicative extensions among the classes τ_i.

COROLLARY 7.10. *Modulo elements killed by inverting a_σ, we have a relation*

$$\tau_i^2 = a_{\bar{\rho}_G} \tau_{i+1}.$$

For computational reasons, it is convenient to extend this result very slightly and kill 2 in the homotopy. Let $\tau_0 \colon S^1 \to H\underline{\mathbb{Z}}/2 \wedge_{BP^{(n)}} H\underline{\mathbb{Z}}/2$ denote the mod 2 Bockstein.

PROPOSITION 7.11. *The image of τ_0 in $\pi_* \Phi^G(H\underline{\mathbb{Z}}/2 \wedge_{BP^{(n)}} H\underline{\mathbb{Z}}/2) = \mathbb{F}_2[\sqrt{b}, \sqrt{b'}]$ is $\sqrt{b'}$.*

COROLLARY 7.12. *for all $i \geq 0$, we have*

$$\tau_i^2 = a_{\bar{\rho}_G} \tau_{i+1}.$$

REMARK 7.13. This result has a number of antecedents. In motivic homotopy theory, we can also form $H\mathbb{F}_2 \wedge_{BP} H\mathbb{F}_2$. Work of Voevodsky shows that the homotopy groups of this are $\pi_\star H\mathbb{F}_2[\tau_0, \ldots]/\tau_i^2 = \rho\tau_{i+1}$, where ρ is a distinguished class in the motivic homotopy of a point.

The computations in this section have been in homology. We turn now to cohomology to produce our desired differentials. Let Q_i denote the class in ($BP^{(n)}$-relative) cohomology dual to τ_i. This is an element in $\pi_\star F_{BP^{(n)}}(H\underline{\mathbb{Z}}/2, H\underline{\mathbb{Z}}/2)$, the ring of operations on mod 2-cohomology in the category of $BP^{(n)}$-modules. Since the dual has a product, this has a coproduct, and the multiplicative relations on the classes τ_i gives us a coproduct relations on the classes Q_i.

PROPOSITION 7.14. *The class $a_{\bar{\rho}_G} Q_{i-1} \otimes Q_{i-1}$ occurs in the coproduct of Q_i with non-zero coefficient.*

These classes are not *a priori* in the equivariant Steenrod algebra. These are classes in the homotopy of the spectrum of $BP^{(n)}$-module endomorphisms of $H\underline{\mathbb{Z}}/2$. However, since the slice filtration is a filtration of $BP^{(n)}$-modules, this is sufficient for our purposes. In fact, though not needed, it is not difficult to show for naturality reasons that these map to non-trivial elements in the equivariant Steenrod algebra.

THEOREM 7.15. *Modulo classes annihilated by inverting* a_σ,

$$Q_i(u_\sigma^{2^i}) = a_{\bar\rho_G}^{2^i-1} a_\sigma^{2^i}.$$

PROOF. The equivariant chain complexes show that $Q_0(u_\sigma) = a_\sigma$. The result now follows by induction. □

THEOREM 7.16. *In the slice spectral sequence for* $BP^{(n)}$, *we have differentials*

$$d(u_{2\sigma}^{2^i}) = a_{\bar\rho_G}^{2^{i+1}-1} a_\sigma^{2^{i+1}} N(\bar v_i).$$

PROOF. This is immediate from the definition of the classes τ_i: τ_i detects multiplication by $N(\bar v_i)$. The previous theorem them gives this result modulo classes annihilated by a_σ. A simple computation shows that in this range of possible targets, multiplication by a_σ is faithful. □

As we shall see, this theorem is the essential result needed for the periodicity and homotopy fixed points theorems.

7.3. Interlude $K_{\mathbb{R}}$ and KO. To ground the previous differentials, we present the example of the C_2-fixed points of $K_{\mathbb{R}}$. This computation was originally done by Dugger with his C_2-slice spectral sequence [**10**]. We reproduce it with our language and machinery. The E_2-page of the spectral sequence is shown in Figure 5. We have included the names of several classes to help the reader. In particular, classes that occur on the zero line are powers of $\bar v_1$ times complementary powers of $u_{2\sigma}$. As always, boxes are copies of $\mathbb{Z}$ while black dots are copies of $\mathbb{Z}/2$.

The class $\bar v_1$ is an actual equivariant homotopy class, so we know it is a permanent cycle in the slice spectral sequence. Similarly, a_σ is in the Hurewicz image. The only class unaccounted for is $u_{2\sigma}$, and Theorem 7.16

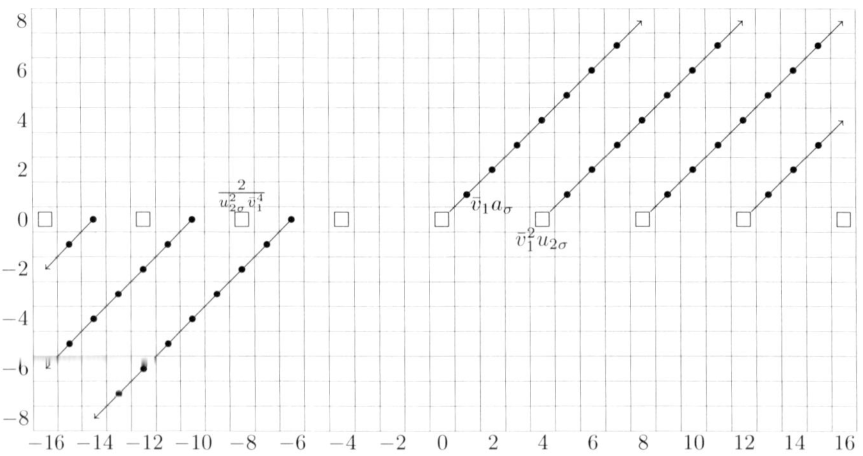

FIGURE 5. The Slice E_2-term for $K_{\mathbb{R}}$

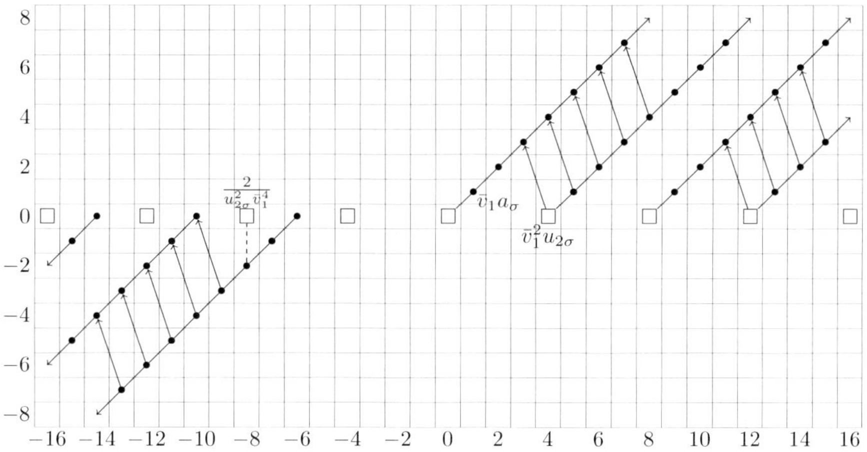

FIGURE 6. The Slice E_3-term for $K_{\mathbb{R}}$

shows that it supports a slice d_3:

$$d_3(u_{2\sigma}) - a_\sigma^3\bar{v}_1.$$

This differential takes place in an equivariant stem (connecting the $2-2\sigma$ and the $1-2\sigma$ stems). However, since this is a spectral sequence of algebras, we see $\mathbb{Z}$-graded versions of this:

$$d_3(\bar{v}_1^2 u_{2\sigma}) = \bar{v}_1^2 d_3(u_{2\sigma}) = a_\sigma^3\bar{v}_1^3.$$

We present the E_3-page together with this differential in Figure 6. The reader is directed to the positive parts of the homotopy, where the slice spectral sequence looks very much like the homotopy fixed points spectral sequence under the line of slope 1 through the origin. In negative degrees, we see essentially the dual pattern, and we will not dwell much on those stems.

For $K_{\mathbb{R}}$, the class $\bar{v}_1$ is a unit. This means that we also have a differential

$$d_3(\bar{v}_1^{-1}u_{2\sigma}) = a_\sigma^3,$$

and so $a_\sigma^3 = 0$ from E_4 on. This produces a horizontal vanishing line in the spectral sequence, and $E_4 = E_\infty$. In particular (even if E_4 were not E_∞), we see that $u_{2\sigma}^2$ is a permanent cycle. We also see that since a_σ is nilpotent, the geometric fixed points of $K_{\mathbb{R}}$ is contractible. All of this followed simply from the invertibility of $\bar{v}_1$, and these types of properties will be key in our identification of the periodicity.

8. Homotopy fixed points, periodicity, and the spectrum Ω

In this section, we will complete our sketch, tying our Gap Theorem with the differentials we discovered. In all cases, the classes $N_{2^k}^{2^n}(\bar{r}_{i,k})$ play a special role: they are carried by regular representation spheres and are seen by the differentials. Thus upon inverting families of these classes, we get spectra with a gap and for which the slice spectral sequence is simplified. Since these norms will occur quite frequently in all that follows, let

$$N_k^n(x) = N_{C_{2^k}}^{C_{2^n}}(x).$$

8.1. The homotopy fixed point and periodicity theorems. The last results needed for various localizations of $MU_{(C_{2^n})}$ rely on an interplay between things we can compute for the homotopy fixed points and things we can compute for honest fixed points. We begin with a slightly general theorem that will greatly help us. Let EG denote the free contractible G-space. Recall also that we defined spaces $E\mathcal{F}_m$ to be the universal space for the family of proper subgroups of C_{p^m}.

THEOREM 8.1. *Let R be a G ring spectrum. If for all $m > 0$ $\big(\tilde{E}\mathcal{F}_m \wedge R\big)^{C_{p^m}}$ is contractible, then the natural map*

$$R \to F(EG_+, R)$$

is a G-weak equivalence.

PROOF. The proof is by induction on the order of the group. If the the group is trivial, then the result is obviously true. Assume now that for all subgroups of C_{p^n} the result is true and that the geometric fixed point object is contractible. If we smash the map $\iota\colon R \to F(EG_+, R)$ with the isotropy separation sequence, then we have a diagram

$$\begin{array}{ccccc} E\mathcal{F}_{n+} \wedge R & \longrightarrow & R & \longrightarrow & \tilde{E}\mathcal{F}_n \wedge R \\ {\scriptstyle E\mathcal{F}_{n+}\wedge\iota}\downarrow & & {\scriptstyle \iota}\downarrow & & {\scriptstyle \tilde{E}\mathcal{F}_n\wedge\iota}\downarrow \\ E\mathcal{F}_{n+} \wedge F(EG_+, R) & \longrightarrow & F(EG_+, R) & \longrightarrow & \tilde{E}\mathcal{F}_n \wedge F(EG_+, R). \end{array}$$

By induction, the map labeled $E\mathcal{F}_{n+} \wedge \iota$ is a G-weak equivalence. By the assumption on R, the upper right corner is equivariantly contractible. The map labeled $\tilde{E}\mathcal{F}_n \wedge \iota$ is a ring map, so we learn that it must also be an equivalence. Therefore the map labeled ι is an equivalence. □

The results of the previous section show us exactly how to ensure that these geometric fixed points sets are contractible for various $MU_{(C_{2^n})}$-algebras.

THEOREM 8.2. *Let R be an $MU_{(C_{2^n})}$-algebra. If $N_1^n(\bar{r}_{2^i-1,k})$ is a unit in $\pi_\star R$, then $\big(\tilde{E}\mathcal{F}_k \wedge R\big)^{C_{2^k}}$ is contractible.*

PROOF. Since we are computing the homotopy groups of the C_{2^k}-fixed points, by naturality of the slice spectral sequence we may assume that $k = n$ and show the result there.

Theorem 7.16 show that we have a slice differential

$$u_{2\sigma_n}^{2^{i-1}} \mapsto a_{\bar{\rho}}^{2^i-1} a_{\sigma}^{2^i} N_2^n(\bar{r}_{2^i-1,n}).$$

Since $N_2^n(\bar{r}_{2^i-1,n})$ is a unit, this allows us to conclude that starting on the next page of the spectral sequence $a_\sigma^{2^i+1}a_{\bar{\rho}}^{2^i-1} = 0$. Simple computations with the complexes occurring in the slice tower show that there are not possible targets for a hidden a_σ-multiplication (as this class is always at the farthest edge of the vanishing region), so we conclude that after inverting a_σ, we get the zero ring. This gives the result. □

Corollary 8.3. *If R is an $MU_{(C_{2^n})}$-algebra in which $N_1^n(\bar{r}_{2^i-1,k})$ is a unit, then $u_{2\sigma_k}^{2^i}$ is a permanent cycle.*

Proof. Again, the statement is one taking place for C_{2^k} fixed points, so we may assume $k = n$. If we smash the slice tower of $MU_{(C_{2^n})}$ with $S^{2^j\sigma}$ and take homotopy, then we get a spectral sequence computing π_* $(\Sigma^{2^j\sigma}MU_{(C_{2^n})})$. Using our complexes, it is easy to see that the last possible differential on $u_{2\sigma}^{2^{j-1}}$ (which indeed is the only possible differential) is the one we found using power operations.

Since under the hypothesis the target of the differential on $u_{2\sigma}^{2^i}$ is obviously already killed, we conclude that $u_{2\sigma}^{2^i}$ is a permanent cycle. □

These two results are the heart of the Homotopy Fixed Point and Periodicity theorem.

Theorem 8.4 (Homotopy Fixed Point and Periodicity Theorems). *Assume given a sequence of non-negative numbers i_1 through i_n, let i be the maximum of the i_j, let i' denote the minimum of the i_j, and assume $\bar{\Delta}\colon S^{m\rho} \to MU_{(C_{2^n})}$ is divisible by $N_1^n(\bar{r}_{2^{i_k}-1,k})$ for all k. Then*

(1) *We have a weak equivalence*

$$\left(\bar{\Delta}^{-1}MU_{(C_{2^n})}\right)^{C_{2^n}} \simeq \left(\bar{\Delta}^{-1}MU_{(C_{2^n})}\right)^{hC_{2^n}}.$$

(2) *The class $u_{2\rho}^{2^i}$ lifts to a non-zero homotopy class in $\bar{\Delta}^{-1}MU_{(C_{2^n})}$.*

(3) *$\pi_*\bar{\Delta}^{-1}MU_{(C_{2^n})}$ is $2^n \cdot lcm(2^{i+1},(2^{i'}-1))$-periodic.*

While not immediately obviously so, all three parts of this result are intimately related and are simple consequences of our earlier theorems. For this reason, we link them into a single result

Proof. The proof of the first part is immediate from Theorems 8.1 and 8.2. For the second part, we know from Corollary 8.3 that for all k, $u_{2\sigma_k}^{2^i}$ is a permanent cycle and therefore a homotopy class. This means that for all k $N_k^n(u_{2\sigma_k}^{2^i})$ is a permanent cycle and therefore a homotopy class, and so multiplying these classes gives

$$u_{2\rho}^{2^i} = \prod_{k=1}^{n} N_k^n(u_{2\sigma_k}^{2^i})$$

is a permanent cycle and a homotopy class.

The third result combines the first two. The class $u_{2\rho}^{2^i}$ gives an equivariant map

$$\Sigma^{2^{i+n+1}}\bar{\Delta}^{-1}MU_{(C_{2^n})} \to \Sigma^{2^{i+1}\rho}\bar{\Delta}^{-1}MU_{(C_{2^n})},$$

and when we restrict to the trivial group, this map is the identity. Thus we have an equivariant map that is an underlying equivalence, and that gives us an equivalence on homotopy fixed points. Since these are the same as honest fixed points, we conclude that the class $u_{2\rho}^{2^i}$ is invertible in $\pi_\star\bar{\Delta}^{-1}MU_{(C_{2^n})}$. The desired periodicity arises by multiplying $u_{2\rho}^{2^i}$ by the appropriate powers of $N_k^n(\bar{r}_{2^{i_k}-1,k})$ as k-varies (all of which will be invertible classes in trivial stems). If we chose i_k to be minimal, then we get the smallest predictable periodicity. □

Since $\bar{\Delta}$ in the previous theorem is carried by a regular representation sphere, we also conclude that $\pi_{-2}\bar{\Delta}^{-1}MU_{(C_{2^n})} = 0$. Since the homotopy groups are periodic, this shows that a large number of groups are actually zero. We should stress here the importance of the Homotopy Fixed Point theorem. The Detection ad Periodicity Theorems are proved for the homotopy fixed points of $\tilde{\Omega}$, not the honest fixed points. On the other hand, the Gap Theorem is proved for honest fixed points which *a priori* only map to the homotopy fixed points. The Homotopy Fixed Point Theorem provides the bridge between the two kinds of results.

8.2. The spectrum Ω and the detection theorem. As was described early on, our spectrum $\tilde{\Omega}$ is a stand-in for E_4 with the C_8-action. We will build $\tilde{\Omega}$ by inverting equivariant classes for each of the subgroups of C_8, just as in the Homotopy Fixed Points Theorem, while ensuring that we have good equivariant maps from these localizations to E_4.

We first look at the underlying spectrum for MU_{C_8}. This is $MU \wedge MU \wedge MU \wedge MU$ with the permutation action of C_8, and the underlying homotopy groups represent the data of having 4 formal group laws $F_1, \ldots, F_4$ together with a chain of isomorphisms

$$F_1 \xrightarrow{\gamma} F_2 \xrightarrow{\gamma} F_3 \xrightarrow{\gamma} F_4$$

such that $\gamma^4 = [-1](x)$. In particular, we have a canonical map

$$\pi_*(MU \wedge MU \wedge MU \wedge MU) \to \pi_* E_4$$

that classifies the universal deformation formal group law carried by E_4, together with its translates under the group C_8. This then gives us by construction a C_8-equivariant map

$$MU \wedge MU \wedge MU \wedge MU \to E_4.$$

Through this map, we will isolate equivariant elements which can be safely inverted without destroying the detection of the Kervaire classes.

In particular, we have the following result, since E_4 is v_4-periodic and since $r_{15,1} \equiv v_4$ modulo the previous classes.

PROPOSITION 8.5. *The map*

$$MU \wedge MU \wedge MU \wedge MU \to E_4$$

factors through $\left(N_1^3\bar{v}_4\right)^{-1} MU \wedge MU \wedge MU \wedge MU$.

To fully spell out the reasoning for inverting the classes we do, we must have a brief digression into the Morava stabilizer group $\mathbb{S}_4$ [**25**]. By definition, this is the automorphism group of the Honda formal group law, and it can be realized as the group of units of the maximal order of the division ring over $\mathbb{Q}_2$ of Hasse invariant $1/4$. In particular, using a presentation of this division ring, we can write any element $g \in \mathbb{S}_4$ as

$$g = \sum_{j=0}^{\infty} t_j(g^{-1}) T^j,$$

where $t_j(g^{-1})$ is an element of $\mathbb{F}_{15}$. As is suggested by the notation, the symbols t_j are functions on the Morava stabilizer group, and they are also the images of the elements by the same name in BP_*BP. Thus this formula reflects the interplay between the BP_*BP-coaction and the group action. For us, the most important part is that we can check if the value of t_i on a group element is a non-zero element of π_*E_4 modulo the maximal ideal. Since π_*E_4 is a complete local ring, any element in π_*E_4 that has non-zero reduction modulo the maximal ideal is a unit. Restricting attention to C_4, this gives us another class we can invert, since an elementary argument involving valuations shows that $t_2(\gamma^2) \neq 0$. Modulo lower classes, $t_2 \equiv r_{3,2}$.

PROPOSITION 8.6. *The map*

$$MU \wedge MU \wedge MU \wedge MU \to E_4$$

factors through $\left(N_1^3\bar{r}_{3,2}\right)^{-1} MU \wedge MU \wedge MU \wedge MU$.

As stated, the result about the interplay between the elements t_i and the group action is for $MU \wedge MU$ (really $BP \wedge BP$). We are working with $MU \wedge MU \wedge MU \wedge MU$. However, a similar analysis shows that we have the same relationship between the classes $r_{2^i-1,3}$ and the generator γ (in fact, this is essentially a way we could define $r_{1,3}$). The same valuation argument use for C_4 shows that $r_{1,3}(\gamma) \neq 0$.

PROPOSITION 8.7. *The map*

$$MU \wedge MU \wedge MU \wedge MU \to E_4$$

factors through $\left(N_1^3\bar{r}_{1,3}\right)^{-1} MU \wedge MU \wedge MU \wedge MU$.

At this point, we can no longer say if there are any elements we can obviously invert. We have inverted appropriate norm classes for each subgroup, and we know that the resulting map has an equivariant map from the underlying spectrum to E_4. We can therefore define $\tilde{\Omega}$.

Definition 8.8. Let $\tilde{\Omega}$ be

$$\left(N_1^3\bar{r}_{1,3}\right)^{-1}\left(N_1^3\bar{r}_{3,2}\right)^{-1}\left(N_1^3\bar{r}_{15,1}\right)^{-1}MU_{C_8}.$$

Recall that Ω was defined to be the fixed points of $\tilde{\Omega}$. An immediate application of our big theorems shows the following.

Corollary 8.9.

(1) *The homotopy groups of* Ω *are* 256*-periodic.*
(2) *We have* $\pi_{-2}\Omega = 0$.
(3) *We have an equivalence* $\Omega \simeq \tilde{\Omega}^{hC_8}$.

These are exactly the properties needed to prove the Kervaire theorem. We state without proof the final piece. The argument for this is classical: we show using the homotopy fixed point spectral sequence and the comparison with the Adams-Novikov spectral sequence that any element representing the Kervaire classes maps to a non-zero element. Since the argument is of a decidedly different flavor, we refer the reader to the main paper.

Theorem 8.10 (Detection Theorem). *If* $\theta_j \in \pi_{2^{j+1}-2}S^0$ *is non-zero, then it has non-zero Hurewicz image in* $\pi_*\Omega$.

Since these elements are then detected in the zero group for $j \geq 7$, we conclude that these elements do not survive.

References

[1] Shôrô Araki. Coefficients of MR-theory. Available online at www.math.rochester.edu/u/faculty/doug/.

[2] Shôrô Araki. Orientations in τ-cohomology theories. *Japan. J. Math. (N.S.)*, 5(2):403–430, 1979.

[3] M. F. Atiyah. K-theory and reality. *Quart. J. Math. Oxford Ser. (2)*, 17:367–386, 1966.

[4] Maia Averett. *Completion of Real Johnson-Wilson theory* $E\mathbb{R}(n)$ *gives fixed points of Morava E-theory.* PhD thesis, UCSD.

[5] G. Carlsson. A survey of equivariant stable homotpy theory. *Topology*, 31:1–28, 1992.

[6] Ethan S. Devinatz, Michael J. Hopkins, and Jeffrey H. Smith. Nilpotence and stable homotopy theory. I. *Ann. of Math. (2)*, 128(2):207–241, 1988.

[7] T. tom Dieck. Kobordismtheorie klassifiziender Räume und Transformationsgrupen. *Mathematische Zeitschrift*, 126:31–39, 1972.

[8] T. tom Dieck. *Transformation Groups and Representation Theory*, volume 766 of *Lecture Notes in Mathematics*. Springer-Verlag, New York, 1979.

[9] Andreas W. M. Dress. Contributions to the theory of induced representations. In *Algebraic K-theory, II: "Classical" algebraic K-theory and connections with arithmetic (Proc. Conf., Battelle Memorial Inst., Seattle, Wash., 1972)*, pages 183–240. Lecture Notes in Math., Vol. 342. Springer, Berlin, 1973.

[10] Daniel Dugger. An Atiyah-Hirzebruch spectral sequence for KR-theory. *K-Theory*, 35(3–4):213–256 (2006), 2005.

[11] Leonard Evens. A generalization of the transfer map in the cohomology of groups. *Trans. Amer. Math. Soc.*, 108:54–65, 1963.

[12] E. Dror Farjoun. *Cellular spaces, null spaces and homotopy localization*, volume 1622 of *Lecture Notes in Mathematics*. Springer-Verlag, New York, 1996.

[13] J. P. C. Greenlees and J. P. May. Equivariant stable homotopy theory. In *Handbook of algebraic topology*, pages 277–323. North-Holland, Amsterdam, 1995.
[14] J. P. C. Greenlees and J. P. May. Localization and completion theorems for MU-module spectra. *Ann. of Math. (2)*, 146(3):509–544, 1997.
[15] Michael A. Hill, Michael J. Hopkins, and Douglas C. Ravenel. On the non-existence of elements of Kervaire invariant one. Available on the arxiv, 2009.
[16] Michael J. Hopkins and Fabien Morel. On the zero slice of MGL and S^0. In preparation.
[17] Po Hu and Igor Kriz. Real-oriented homotopy theory and an analogue of the Adams-Novikov spectral sequence. *Topology*, 40(2):317–399, 2001.
[18] Nitu Kitchloo and W. Stephen Wilson. On the Hopf ring for $ER(n)$. *Topology Appl.*, 154(8):1608–1640, 2007.
[19] Peter S. Landweber. Conjugations on complex manifolds and equivariant homotopy of MU. *Bull. Amer. Math. Soc.*, 74:271–274, 1968.
[20] L. G. Lewis, J. P. May, and M. Steinberger. *Equivariant Stable Homotopy Theory*, volume 1213 of *Lecture Notes in Mathematics*. Springer-Verlag, New York, 1986.
[21] L. Gaunce Lewis, Jr. The $RO(G)$-graded equivariant ordinary cohomology of complex projective spaces with linear $\mathbf{Z}/p$ actions. In *Algebraic topology and transformation groups (Göttingen, 1987)*, volume 1361 of *Lecture Notes in Math.*, pages 53–122. Springer, Berlin, 1988.
[22] M. A. Mandell and J. P. May. Equivariant orthogonal spectra and S-modules. *Mem. Amer. Math. Soc.*, 159(755):x+108, 2002.
[23] M. A. Mandell, J. P. May, S. Schwede, and B. Shipley. Model categories of diagram spectra. *Proc. London Math. Soc. (3)*, 82(2):441–512, 2001.
[24] J. P. May. *Equivariant homotopy and cohomology theory*, volume 91 of *CBMS Regional Conference Series in Mathematics*. Published for the Conference Board of the Mathematical Sciences, Washington, DC, 1996. With contributions by M. Cole, G. Comezaña, S. Costenoble, A. D. Elmendorf, J. P. C. Greenlees, L. G. Lewis, Jr., R. J. Piacenza, G. Triantafillou, and S. Waner.
[25] Douglas C. Ravenel. *Complex cobordism and stable homotopy groups of spheres*, volume 121 of *Pure and Applied Mathematics*. Academic Press Inc., Orlando, FL, 1986. Errata and second edition available online at author's home page.
[26] Charles Rezk. Notes on the Hopkins-Miller theorem. In *Homotopy theory via algebraic geometry and group representations (Evanston, IL, 1997)*, volume 220 of *Contemp. Math.*, pages 313–366. Amer. Math. Soc., Providence, RI, 1998.
[27] V. Voevodsky. On the zero slice of the sphere spectrum. *Tr. Mat. Inst. Steklova*, 246(Algebr. Geom. Metody, Svyazi i Prilozh.):106–115, 2004.
[28] Vladimir Voevodsky. Open problems in the motivic stable homotopy theory. I. In *Motives, polylogarithms and Hodge theory, Part I (Irvine, CA, 1998)*, volume 3 of *Int. Press Lect. Ser.*, pages 3–34. Int. Press, Somerville, MA, 2002.
[29] Vladimir Voevodsky. A possible new approach to the motivic spectral sequence for algebraic K-theory. In *Recent progress in homotopy theory (Baltimore, MD, 2000)*, volume 293 of *Contemp. Math.*, pages 371–379. Amer. Math. Soc., Providence, RI, 2002.
[30] Klaus Wirthmüller. Equivariant homology and duality. *Manuscripta Math.*, 11: 373–390, 1974.
[31] Michael A. Hill, Michael J. Hopkins, and Douglas C. Ravenel. The Arf-Kervaire invariant problem in algebraic topology: introduction. In *Current developments in mathematics, 2009*, pages 23–57, Int. Press, Somerville, MA, 2010.

Current Developments in Mathematics, 2010

On the Friedlander-Milnor conjecture for groups of small rank

Fabien Morel

Contents

0. Introduction

The title refers to the following conjecture of E. Friedlander [**12**]:

CONJECTURE 1. *For any algebraically closed field F, any reductive algebraic F-group G and any prime ℓ different from $char(F)$ the natural ring homomorphism*

$$H^*_{et}(BG;\mathbb{Z}/\ell) \to H^*(BG(F);\mathbb{Z}/\ell)$$

from the $\mathbb{Z}/\ell$-étale cohomology of the simplicial classifying space of G to the cohomology with $\mathbb{Z}/\ell$-coefficients of the discrete group $G(F)$ is an isomorphism.

and with J. Milnor's variant in the case $F = \mathbb{C}$ [**35**]:

CONJECTURE 2. *For any Lie group G and any prime ℓ the natural ring homomorphism*

$$H^*_{sing}(BG;\mathbb{Z}/\ell) \to H^*(BG^\delta;\mathbb{Z}/\ell)$$

from the $\mathbb{Z}/\ell$-singular cohomology of the classifying space BG of the topological group G to the $\mathbb{Z}/\ell$-singular cohomology of the classifying space BG^δ of the discrete group G^δ underlying G is an isomorphism.

The two conjectures are equivalent for $F = \mathbb{C}$ and G a complex algebraic Lie group by Grothendieck's comparisons theorems between étale and singular cohomology, but each contains cases which are not reached by the other.

The aim of these conjectures is the computation of the right hand side that is to say the mod ℓ cohomology of the (classifying space of the) discrete group $G(F)$. Indeed, again by Grothendieck's comparisons theorems the left hand side of conjecture 1 is isomorphic to the singular mod ℓ cohomology of the topological classifying space of the complex algebraic Lie group $G(\mathbb{C})$ of complex points of G. The singular cohomology of the classifying space of a Lie group is in practice already computed (in most cases) by standard technics of algebraic topology and should be considered as "simpler". For instance $H^*_{sing}(BSL_n(\mathbb{C});\mathbb{Z}/\ell)$ is known to be a polynomial ring with coefficients in $\mathbb{Z}/\ell$ on the Chern classes $c_2,\ldots,c_n$.

The main evidence for these conjectures is the result of Suslin [**60, 61**] which establish the case $G = SL_\infty$ (which is not an algebraic group!) for each separably closed field F. Friedlander and Mislin also proved the conjecture for G arbitrary and F the algebraic closure of a finite field [**14**]. Using Suslin's results [**62**] on stability properties of the homology for SL_n one may deduce the conjecture for SL_n in (co-)homological degree $*$ small enough compared to n. For the classical groups like the orthogonal and symplectic group, one may as well deduce from Suslin Theorem partial results using stability properties; see for instance [**29**]. Except these cases obtained from Suslin's result by stabilization properties, the conjecture is basically unknown. For further references the reader should check [**30**] and [**50**]. Up to now in particular, there is no known split simple algebraic group G for which the conjecture holds over any separably closed field F.

Let us start with the following result which establishes what we call "the weak Friedlander-Milnor conjecture":

THEOREM 3. *Let F be a separably closed field, let G be any reductive algebraic F-group G and let ℓ be a prime different from $char(F)$. Then the natural homomorphism*

$$H^*_{et}(BG;\mathbb{Z}/\ell) \to H^*(B(G(\Delta^\bullet_F));\mathbb{Z}/\ell)$$

is an isomorphism.

Here $\Delta^\bullet_F$ denotes the cosimplicial affine F-scheme defined by $\Delta^n_F := Spec(F[T_0,\dots,T_n]/(\sum_i T_i - 1))$ and with the standard cosimplicial structure. In the previous Theorem $G(\Delta^\bullet_F)$ denotes the simplicial group obtained by evaluating G on $\Delta^\bullet_F$, and $BG(\Delta^\bullet_F)$ means its classifying space (as in [**9**] for instance). Thus the right hand side of the previous isomorphism is the mod ℓ singular cohomology of classifying space of the simplicial group $G(\Delta^\bullet_F)$. The morphism itself comes from a commutative square of the form

$$\begin{array}{ccc} H^*_{et}(BSing^{\mathbb{A}^1}_\bullet(G));\mathbb{Z}/\ell) & \to & H^*(B(G(\Delta^\bullet_F));\mathbb{Z}/\ell) \\ \downarrow\wr & & \downarrow \\ H^*_{et}(BG;\mathbb{Z}/\ell) & \to & H^*(B(G(F));\mathbb{Z}/\ell) \end{array}$$

in which $Sing^{\mathbb{A}^1}_\bullet(G)$ is the Suslin-Voevodsky construction on G (see below), the left vertical morphism is an isomorphism and the bottom horizontal morphism is the one appearing in the Friedlander conjecture. This diagram will be discussed in Section 1.1 below.

REMARK 4. Classical devissage methods and Theorems for the structure of algebraic groups over perfect fields allow one to extends the previous Theorem to smooth F-groups scheme which can be defined over a perfect subfield of F. The cases where G is an abelian variety, a smooth unipotent group defined over a perfect subfield, a discrete finite group, or a split Torus are straightforward to check. Thus the main point of the Theorem and of the Friedlander-Milnor conjecture is the case where G is (semi-)simple. □

The following property depending on the smooth F-group G plays an important role in relating Theorem 3 to the Friedlander conjecture through the above diagram:

DEFINITION 5. Assume F is an infinite field. Let G be a smooth algebraic F-group and let ℓ be any prime different from $char(F)$. One says that G satisfies *the homotopy invariance of group cohomology at ℓ* if for any integer $n \geq 1$ the natural ring homomorphism

$$H^*(BG(F[T_1,\dots,T_n]);\mathbb{Z}/\ell) \to H^*(BG(F);\mathbb{Z}/\ell)$$

is an isomorphism.

We observe (see [**30**, page 110]) that if F is finite the homotopy invariance of group cohomology is false in general. The previous notion is relevant because of the following consequence:

COROLLARY 6. *Let F be a separably closed field, let G be any reductive algebraic F-group and ℓ be a prime different from char(F). If G satisfies the homotopy invariance of group cohomology at ℓ then the map of simplicial sets:*

$$BG(F) \to B(G(\Delta^\bullet_F))$$

induces an isomorphism on mod ℓ-cohomology and consequently the Friedlander conjecture holds for G.

PROOF. This follows by using the spectral sequences of the bisimplicial set $BG(\Delta^\bullet_F)$ of the form

$$E_1^{p,q} = H^q(BG(\Delta^p_F); \mathbb{Z}/\ell) \Rightarrow H^{p+q}(BG(\Delta^\bullet_F); \mathbb{Z}/\ell)$$

together with the fact that each of the F-scheme Δ^p_F is an affine F-space of dimension p. □

From [**24**] it follows that the group SL_2 doesn't have the property of homotopy invariance of group cohomology at ℓ. The weak Friedlander-Milnor conjecture reduces the Friedlander-Milnor conjecture for SL_2 to the statement that the morphism

$$BSL_2(F) \to B(SL_2(\Delta^\bullet_F)) \tag{0.1}$$

induces an isomorphism on mod ℓ homology for any separably closed field F of characteristic $\neq \ell$.

The SL_2 case should be quite special and we believe that each semi-simple group, not containing the type SL_2, should satisfy the homotopy invariance of group cohomology at ℓ, at least over a separably closed field F of characteristic $\neq \ell$.

For the group SL_3 the homotopy invariance for group cohomology is proven in [**30**, Theorem 4.6.8] because for polynomial rings with coefficients in a field F one knows that E_3 agrees with SL_3 by results of Suslin [**59**]. The Theorem of Knudson is an elaboration on the work [**56**] of Soulé who proved the invariance property for each split semi-simple groups in the case of polynomial rings in one variable. In [**70**] the technics of Knudson are generalized to any split semi-simple group of rank 2. Moreover Wendt also proved in [**71**] that SL_4 has the homotopy invariance of group cohomology at ℓ. Consequently we get altogether:

THEOREM 7. *Let F be a separably closed field and let ℓ be a prime different from char(F). Let G be a split simple algebraic F-group of one of the following type: SL_3, SO_5 or G_2 and SL_4. Then the natural homomorphism*

$$H^*_{et}(BG; \mathbb{Z}/\ell) \to H^*(BG(F); \mathbb{Z}/\ell)$$

is an isomorphism.

REMARK 8. In a work still in progress, we hope to fix the case of SL_2 as well by establishing that (0.1) induces an isomorphism on mod ℓ homology for any separably closed field F of characteristic $\neq \ell$. □

When F is the field of complex numbers, Grothendieck's comparison theorems give some cases of Milnor's form of the conjecture for complex algebraic Lie groups:

COROLLARY 9. *For any complex algebraic Lie group G and any prime ℓ the natural ring homomorphism*

$$H^*_{sing}(BG;\mathbb{Z}/\ell) \to H^*(BG(\Delta^\bullet_{\mathbb{C}})^\delta;\mathbb{Z}/\ell)$$

is an isomorphism. Here $G(\Delta^n_{\mathbb{C}})^\delta$ means the discrete group of algebraic morphisms $\Delta^n_{\mathbb{C}} \to G$, and BG means the classifying space of the topological group underlying G. Consequently the morphism

$$H^*(BG;\mathbb{Z}/\ell) \to H^*(BG^\delta;\mathbb{Z}/\ell)$$

is an isomorphism if G is a complex simple algebraic group of type SL_3, SO_5, G_2 or SL_4.

In Section A of the Appendix we give a direct treatment of the degree 2 case of the Friedlander conjecture, which relies on standard facts on central extensions of algebraic groups and the theory of $\mathbb{A}^1$-universal coverings developed in [**41**]. This should be compared to the results of [**50**]. Using a slight elaboration of the technics of this work, one may in fact go one step further and prove the conjecture for $* \leq 3$ for any group.

To explain our strategy for the weak Friedlander conjecture, *alias* Theorem 3, let us recall that the Quillen algebraic K-theory groups $K_n(F)$ of a field F (or of a regular ring) are isomorphic in degree $n \geq 1$ to the Karoubi-Villamayor K-theory groups $K_n^{KV}(F) := \pi_n(B(\mathbb{G}L_\infty(\Delta^\bullet_F))$, see for instance [**17**]. The argument of Suslin relies on the existence of certain transfers morphisms in mod ℓ algebraic K-theory groups, which are used in a fundamental way to prove the famous rigidity theorem.

Our approach will use extensively the Suslin-Voevodsky construction on G, denoted by $Sing^{\mathbb{A}^1}_\bullet(G)$, that is to say the previous construction, but really considered as a simplicial sheaf of groups

$$Sm_k \ni X \mapsto Sing^{\mathbb{A}^1}_\bullet(G)(X) := G(\Delta^\bullet_k \times X)$$

on the category Sm_k of smooth k-schemes in the Nisnevich topology, where k is a fixed perfect base field (like a prime field for instance). Recall [**47, 44**] that the Nisnevich topology has for covering families $\{U_i \to X\}_i$ in Sm_k the étal covering families with the property that for any point $x \in X$ there is an i and a point y in U_i lying over x with the same residue field. Except otherwise stated, in this text we only use the Nisnevich topology on Sm_k.

To be more precise on our approach, we will study the "$\mathbb{A}^1$-homotopy type" of its classifying space $BSing^{\mathbb{A}^1}_\bullet(G)$, a simplicial sheaf of sets on Sm_k in the Nisnevich topology, or a *space* in short, see [**44**]. The main obstruction to adapt Suslin's method to the space $BSing_\bullet^{\mathbb{A}^1}(G)$ comes from the fact that the correspondence $F \mapsto \pi_n(BSing_\bullet^{\mathbb{A}^1}(G)(F))$ doesn't form a cohomology theory and that there is no way, *a priori*, to endow it with suitable transfers, as opposed to the above case $G = SL_\infty$.

To get round this difficulty we will introduce a rather new approach based on $\mathbb{A}^1$-homotopy theory which proceeds roughly speaking in two fundamental steps.

The first step reduces the Theorem 2 to checking a rigidity property for the $\mathbb{A}^1$-chain complexes $C_*^{\mathbb{A}^1}((\mathbb{G}_m)^{\wedge n};\mathbb{Z})$ of the smash-powers of $\mathbb{G}_m$'s for all n's. This reduction step is highly of homotopical nature, but also uses in an essential the Bruhat decomposition for split groups, and also relies essentially on some highly non trivial results from [**41**]. These results, and part of their proof, contains, on the other hand, the "genes" of the notion of sheaves with generalized transfers which is used in the second step.

Observe that as a consequence, the first step reduces Theorem 2 to a statement which doesn't depend on any group G, nor any field F.

The second step is to prove the expected rigidity property for these chain complexes $C_*^{\mathbb{A}^1}((\mathbb{G}_m)^{\wedge n};\mathbb{Z})$: one has to understand why applying the "abstract" $\mathbb{A}^1$-localization functor for chain complexes to $\mathbb{Z}((\mathbb{G}_m)^{\wedge n})$ yields a chain complex whose homology sheaves have the rigidity property at ℓ; this is done by observing that these homology sheaves are endowed with a structure of *sheaves with generalized transfers*, for which the one considered by Voevodsky [**67**] are one of the examples. We conclude by a reduction at the very end to the classical rigidity Theorem of [**63**].

Let us give more details. From now on k is a fixed perfect field.

In Section 1 we give some recollection on the characterization of $\mathbb{A}^1$-local spaces [**44, 41**] in terms of $\mathbb{A}^1$-homotopy sheaves; this is one of the result that we use from [**41**]. We then prove that the classifying space $BSing_\bullet^{\mathbb{A}^1}(G)$ is $\mathbb{A}^1$-local for any split semi-simple group. It is moreover simply connected if G is simply-connected in the sense of algebraic group theory.

In section 2 we observe along the classical lines, see [**60, 63, 28, 30**], that Theorem 3 is a consequence of the "rigidity property at ℓ" of the space $BSing_\bullet^{\mathbb{A}^1}(G)$. It is also well-known (see Knudson's book [**30**] or Jardine's article [**28**]) that this rigidity property for the space BG itself implies directly Friedlander's conjecture. However, though there is so far no known method to prove it for BG directly, in contrast, we are able to prove it for $BSing_\bullet^{\mathbb{A}^1}(G)$. The latter space is a bit more flexible than BG, it should be thought of as the $+$-construction in the $\mathbb{A}^1$-homotopy sense, and can thus be reached by standard method of algebraic topology, in contrast to BG.

To prove the rigidity property at ℓ for $BSing_\bullet^{\mathbb{A}^1}(G)$ we will use the following Theorem which we prove in Section 3:

THEOREM 10. *Let $\mathcal{X}$ be a pointed simply-connected $\mathbb{A}^1$-local space over the base field k, ℓ be a prime different from char(k). If the $\mathbb{A}^1$-chain complexes $C_*^{\mathbb{A}^1}(\mathcal{X}^n;\mathbb{Z}/\ell)$, $n \geq 1$, have the rigidity property at ℓ, then the space $\mathcal{X}$ satisfies the rigidity property at ℓ.*

To prove the Theorem 10, we implement first in Section 3.1 the $\mathbb{A}^1$-lower central series, which is constructed by analogy with [**8, 9**] from the classical lower-central series filtration of the loop space. This gives a canonical expression of the pointed simply connected $\mathbb{A}^1$-local space $\mathcal{X}$ as the homotopy inverse limit of a tower of simply connected pointed spaces

$$\cdots \to W_s(\mathcal{X}) \to W_{s-1}(\mathcal{X}) \to \cdots \to W_1(\mathcal{X}) \to *$$

in which the homotopy fiber of $W_s(\mathcal{X}) \to W_{s-1}(\mathcal{X})$ is the generalized Eilenberg-MacLane space $B(L_{s-1}(KC_*(\mathcal{X})[-1]))$, whose loop space is the free Lie algebra functor L_{s-1} in weight $s-1$ applied to the loop space of the Eilenberg-MacLane space $KC_*(\mathcal{X})$ associated to the chain complex $C_*(\mathcal{X})$ of $\mathcal{X}$. Using [**8**] and the fundamental connectivity results from [**41**], we see that

$$L_{s-1}(KC_*(\mathcal{X})[-1])$$

gets highly and highly $\mathbb{A}^1$-connected as s increases.

The proof of Theorem 10 follows from this in Section 3.2 through a devissage of the total $\mathbb{A}^1$-derived functor of the free Lie algebra functors L_s, that is to say of $\mathbb{A}^1$-localization of spaces of the form $L_{s-1}(KC_*)$, for C_* a complex of abelian sheaves (see [**11, 10**] for the notion of derived functors of non additive functors in the classical case).

We prove that the $\mathbb{A}^1$-chain complexes $C_*^{\mathbb{A}^1}((BSing_\bullet^{\mathbb{A}^1}(G))^n;\mathbb{Z}/\ell)$, $n \geq 1$, have the rigidity property at ℓ in Sections 4 and 5. Using general simple facts on the $\mathbb{A}^1$-chain complexes, we show in Section 4 that the Bruhat decomposition for G reduces the rigidity property at ℓ for the space $BSing_\bullet^{\mathbb{A}^1}(G)$ to the rigidity property of the $\mathbb{A}^1$-chain complexes $C_*^{\mathbb{A}^1}((\mathbb{G}_m)^{\wedge n})$ of smash-products of $\mathbb{G}_m$'s. This finishes the first step of our proof.

The rigidity property of the $\mathbb{A}^1$-chain complexes $C_*^{\mathbb{A}^1}((\mathbb{G}_m)^{\wedge n})$ is proven in section 5. To do this we embed the free sheaf of abelian groups

$$\mathcal{Z}(n) := \mathbb{Z}((\mathbb{G}_m)^{\wedge n})$$

on the n-th smash-power $(\mathbb{G}_m)^{\wedge n}$ of $\mathbb{G}_m$ into an explicit sheaf $\mathcal{Z}_{tr}(n)$ called the transfer completion of $\mathcal{Z}_n$, such that by construction, for any finite field extension $K \subset L$ with a generator $x \in L$, there is a canonical transfer morphism

$$Tr_K^L : \mathcal{Z}(n)(L) \to \mathcal{Z}_{tr}(n)(K)$$

This sheaf $\mathcal{Z}_{tr}(n)$ must be thought of as the sheaf of finite correspondences to $\mathbb{A}^1$ with coefficients in $\mathcal{Z}(n)$, in the spirit of [**67**].

We then prove in a very explicit way that the monomorphism

$$\mathcal{Z}(n) \subset \mathcal{Z}_{tr}(n)$$

is an $\mathbb{A}^1$-quasi-isomorphism, that is to say that the $\mathbb{A}^1$-localization functor $L^{ab}_{\mathbb{A}^1}(-)$ for chain complexes induces a quasi-isomorphism of chain complexes of abelian groups on Sm_k

$$C_*^{\mathbb{A}^1}((\mathbb{G}_m)^{\wedge n}) = L^{ab}_{\mathbb{A}^1}(\mathcal{Z}(n)) \cong L^{ab}_{\mathbb{A}^1}(\mathcal{Z}_{tr}(n))$$

Some general fact on $\mathbb{A}^1$-localization and base change imply that the above transfers morphisms from $\mathcal{Z}(n)(L)$ to $\mathcal{Z}_{tr}(n)(K)$ induce then actual transfers morphisms on the homology sheaves of the chain complex $C_*^{\mathbb{A}^1}((\mathbb{G}_m)^{\wedge n})$.

The $\mathbb{A}^1$-homology sheaves of the $\mathbb{A}^1$-localizations are known [**38, 41**] to be strictly $\mathbb{A}^1$-invariant sheaves. A strictly $\mathbb{A}^1$-invariant sheaf M is a sheaf of abelian groups on Sm_k such that for any $X \in Sm_k$ the map

$$H^*(X; M) \to H^*(\mathbb{A}^1 \times X; M)$$

is an isomorphism. Here the cohomology is computed in the Nisnevich topology. These type of sheaves form an abelian category, and the structure of transfers on those that we obtained is a new one, which we call *strictly $\mathbb{A}^1$-invariants sheaves with generalized transfers*, see Section 5.2 for a precise definition. To be slighlty more precise, this structure involves a structure of sheaf of modules over the sheaf $\underline{\mathbf{GW}} = \underline{\mathbf{K}}_0^{MW}$ of unramified Grothendieck-Witt rings of symmetric non-degenerate bilinear forms constructed in [**41**] (it is called there the Milnor-Witt K-theory ring in weight 0), which is compatible in a tricky sense with the transfers. The structure of homotopy invariant sheaves with transfers defined by Voevodsky [**67**] is exactly the particular case where the $\underline{\mathbf{GW}}$-module structure is trivial, that is to say factors through the rank homomorphism $\underline{\mathbf{GW}} \to \mathbb{Z}$.

We conclude the proof of the rigidity property for the $\mathbb{A}^1$-chain complexes $C_*^{\mathbb{A}^1}((\mathbb{G}_m)^{\wedge n})$ by observing first that for ℓ odd, when -1 is a square in the base field, the $\underline{\mathbf{GW}}$-module structure on the mod ℓ homology sheaves has to be trivial, thus giving homotopy invariant sheaves (of $\mathbb{Z}/\ell$-vector spaces) with transfers in the classical sense of Voevodsky. These sheaves have the rigidity property by [**63**]. Theorem 3 for odd ℓ follows from this.

To prove the rigidity property at the prime $\ell = 2$ requires a bit more work. Roughly speaking we observe that for a strictly $\mathbb{A}^1$-invariant sheaf M with generalized transfers, one may define a filtration

$$\cdots \subset \mathcal{I}^n M \subset \cdots \subset \mathcal{I} M \subset M$$

by strictly $\mathbb{A}^1$-invariant sheaves stable under the transfers and the $\underline{\mathbf{GW}}$-module structures which is directly induced by the filtration of $\underline{\mathbf{GW}}$ by the powers of the fundamental ideal $\underline{\mathbf{I}} \subset \underline{\mathbf{GW}}$. The successive subquotients $\mathcal{I}^n M/\mathcal{I}^{n+1} M$ are thus homotopy invariant sheaves with transfers in the sense of Voevodsky and thus satisfy the rigidity property at 2 as well.

To conclude in the case $\ell = 2$ using this filtration, we observe that for a strict henselization A of a (geometric) point in a smooth k-scheme with fraction field F one has $\mathcal{I}^n M(F) = 0$ for $n \geq 2$. One thus concludes from the rigidity property of $\mathcal{I}^n M/\mathcal{I}^{n+1} M$ for $n = 0$ and 1.

We also observe that just the existence of the subsheaf $\mathcal{I}M \subset M$ is enough to prove the rigidity property at 2 in the original sense of Suslin, that the mod 2 homology $H_*(BSing_\bullet^{\mathbb{A}^1}(G)(F);\mathbb{Z}/2)$ doesn't depend on the separably closed field F. In finite characteristic this suffices to prove Theorem 3 by the results of [**14**]. □

REMARK 11. Using the affirmation by Voevodsky of the Milnor conjectures, which implies (see [**2**, Lemma 2.1]) that for a field F of finite type over its prime field, $I^{n+1}(F) = 2I^n(F)$ for n large enough, we get a much stronger rigidity property, which holds for any henselization of a point of a smooth k-scheme. □

The property that we actually can prove for the space $BSing_\bullet^{\mathbb{A}^1}(G)$ is indeed much stronger that just to imply the weak Friedlander conjecture. It implies also in case -1 is a square in k that the canonical morphism

$$BSing_\bullet^{\mathbb{A}^1}(G) \to (BSing_\bullet^{\mathbb{A}^1}(G))^\wedge_\mu$$

from the space to its motivic completion, obtained either by the method of Sullivan [**58**] or directly as the Bousfield-Kan Tot construction [**5**] applied to the cosimplicial space obtained by using the pair of adjoint functors "Voevodsky motive" and "motivic Eilenberg-MacLane space", is a weak-equivalence when evaluated at fields (or Henselian local rings). We will come back to this point in a sequel to this work. One says then that $BSing_\bullet^{\mathbb{A}^1}(G)$ is *motivic complete.*

To emphasize our method, let us mention the following result which is a generalization of Theorem 3 (obtained by taking for k an algebraic closure of a prime field and $\mathcal{X} = BSing_\bullet^{\mathbb{A}^1}(G)$):

THEOREM 12. *Assume k is an algebraically closed field, ℓ is a prime number different from $char(k)$ and let $\mathcal{X}$ be a pointed $\mathbb{A}^1$-local simply connected space over k. If the $\mathbb{A}^1$-chain complex*[1] *of $\mathcal{X}$ is of mixed Tate type*[2], *then $\mathcal{X}$ is motivic complete, and in particular has the rigidity property at ℓ (see Section 2.1).*

In particular, for any extension $k \subset F$ with F separably closed, the morphism of simplicial sets $\mathcal{X}(k) \to \mathcal{X}(F)$ induces an isomorphism

$$H^*(\mathcal{X}(F);\mathbb{Z}/\ell) \cong H^*(\mathcal{X}(k);\mathbb{Z}/\ell)$$

and moreover the natural ring homomorphism

$$H^*_{et}(\mathcal{X};\mathbb{Z}/\ell) \to H^*(\mathcal{X}(F);\mathbb{Z}/\ell)$$

from the $\mathbb{Z}/\ell$-étale cohomology of the space $\mathcal{X}$ to the mod ℓ cohomology of $\mathcal{X}(F)$ is an isomorphism.

[1] see [**39**][**41**]

[2] see Section 4.1, this basically means that it is build out of $C_*^{\mathbb{A}^1}(\mathbb{G}_m^{\wedge n})$'s

REMARK 13. A further consequence of this result is that, when k is algebraically closed and G is a semi-simple k-group, the morphism of pro-ℓ-simplicial sets

$$(BSing_\bullet^{\mathbb{A}^1}(G)(k))\hat{}_\ell \to (BSing_\bullet^{\mathbb{A}^1}(G))\hat{}_{\ell,et} = (BG)\hat{}_{\ell,et}$$

and also

$$(Sing_\bullet^{\mathbb{A}^1}(G)(k))\hat{}_\ell \to (Sing_\bullet^{\mathbb{A}^1}(G))\hat{}_{\ell,et} = (G)\hat{}_{\ell,et}$$

obtained by applying the pro-ℓ étale homotopy type functor (see for instance [**3, 13**]) to the inclusions of spaces $BSing_\bullet^{\mathbb{A}^1}(G)(k) \subset BSing_\bullet^{\mathbb{A}^1}(G)$ or $Sing_\bullet^{\mathbb{A}^1}(G)(k) \subset Sing_\bullet^{\mathbb{A}^1}(G)$ is an equivalence. Our technic may be used to extend these results to the Suslin-Voevodsky construction on other smooth k-schemes. For instance one gets a pro-ℓ equivalence of the form

$$Sing_\bullet^{\mathbb{A}^1}(\mathbb{P}^n)(k)\hat{}_\ell \cong (\mathbb{P}^n)\hat{}_{\ell,et}$$

by applying the pro-ℓ étale homotopy type functor to the inclusion of spaces $Sing_\bullet^{\mathbb{A}^1}(\mathbb{P}^n)(k)) \subset Sing_\bullet^{\mathbb{A}^1}(\mathbb{P}^n)$. More generally one may derive such equivalences for the Suslin-Voevodsky construction $Sing_\bullet^{\mathbb{A}^1}(X)$ on various homogeneous varieties of the form $X := G/H$, where H is a k-subgroup of G.

This proves that under the previous assumptions, the "very naive" mod ℓ homology of the simplicial set $Sing_\bullet^{\mathbb{A}^1}(X)(k)$ coincides with the mod ℓ Suslin homology of X as it follows from [**63**]. This result is also new. □

Conventions. For a field F, we let Sm_F denote the category of finite type smooth F-schemes.

In the sequel, k denotes a fixed perfect field, G a fixed split semi-simple k-group, and ℓ denotes a prime number different from $char(k)$. We let $\mathcal{F}_k$ denote the category of fields extensions of k which are of finite transcendence degree over k.

If not otherwise stated, by *sheaf* we always mean a sheaf in the Nisnevich topology on Sm_k and by a *space* we mean a simplicial sheaf of sets in the Nisnevich topology on Sm_k.

We refer the reader to [**44, 41**] for the basic notions concerning the simplicial and $\mathbb{A}^1$-homotopy theory of sheaves, the notion of chain complexes and their corresponding Eilenberg-MacLane spaces, and the notions of $\mathbb{A}^1$-homotopy sheaves, $\mathbb{A}^1$-homology sheaves and $\mathbb{A}^1$-chain complexes, which we will use freely.

We will finally assume the reader is acquainted with the notion of essentially smooth k-schemes, these are for us noetherian k-schemes obtained as inverse limit of a system of smooth k-schemes in which the transitions morphisms are étale and affine. For instance the spectrum of the local ring $\mathcal{O}_{X,x}$ of a smooth k-scheme X at a point x. Or its henselization $\mathcal{O}^h_{X,x}$. For such an essentially smooth k-scheme $U = lim_\alpha U_\alpha$, and for $\mathcal{X}$ a space, we will denote by $\mathcal{X}(U)$ the simplicial set obtained as the filtering colimit of the $\mathcal{X}(U_\alpha)$'s. For instance, if $F \in \mathcal{F}_k$, $Spec(F)$ is essentially smooth over k, and $\mathcal{X}(F)$ is the stalk of $\mathcal{X}$ at F.

Acknowledgement. This work is part of a systematic study of the $\mathbb{A}^1$-homotopy type of classifying spaces of algebraic groups started years ago, aiming at understanding the Serre conjectures in Galois cohomology from the point of view of $\mathbb{A}^1$-homotopy theory. I am grateful to Aravind Asok for mentioning to me at some point the Friedlander conjecture. I want to thank Vladimir Voevodsky for his interest during my stay at the Institute for Advanced Studies in 2010. I also want to specially thank Mike Hopkins for his interest in this work, for his support and for some helpful discussions.

The final idea which led to a very important simplification of a previous construction, related to the construction of some transfers morphisms on the $\mathbb{A}^1$-chain complexes of tori, is born at the I.A.S. in the spring 2010. I am very gateful to the I.A.S. for his hospitality during this period January-May 2010, which allowed me to improved considerably the present work in many aspects.

1. The $\mathbb{A}^1$-local space $BSing_\bullet^{\mathbb{A}^1}(G)$

1.1. The Suslin-Voevodsky construction. For $\mathcal{M} : (Sm_k)^{op} \to Set$ a presheaf of sets on the category of smooth k-schemes, we may consider the simplicial presheaf of sets on Sm_k defined by

$$U \mapsto \mathcal{M}(\Delta^\bullet \times U)$$

where $n \mapsto \Delta^n_k = \Delta^n := Spec(k[T_0, \ldots, T_n]/\sum T_i = 1)$ is the standard algebraic cosimplicial simplex over F [**63**]. If $\mathcal{M}$ is a sheaf (in the Nisnevich topology) on Sm_k, this simplicial presheaf is a simplicial sheaf of sets, which we denote by $Sing_\bullet^{\mathbb{A}^1}(\mathcal{M})$. We will call this construction the Suslin-Voevodsky construction on $\mathcal{M}$.

We will be interested in the Suslin-Voevodsky construction $Sing_\bullet^{\mathbb{A}^1}(G)$ on G itself. It is a simplicial sheaf of groups on Sm_k, in fact a sheaf in the étale topology as well, but at this point we will ignore it. We may consider its simplicial classifying space $BSing_\bullet^{\mathbb{A}^1}(G)$ defined by the usual formulas (see for instance [**44**, page 128]). It is a simplicial sheaf of sets on Sm_k and we have a canonical inclusion of simplicial sheaves of sets (coming from the inclusion $G \subset Sing_\bullet^{\mathbb{A}^1}(G)$ as the sheaf of 0-vertices)

$$BG \to BSing_\bullet^{\mathbb{A}^1}(G)$$

Let ℓ be a prime number different from $char(F)$ and let M be a locally constant sheaf of $\mathbb{Z}/\ell$-torsion abelian groups on Sm_F; the classical $\mathbb{A}^1$-invariance property of $H^*_{et}(-; M)$ [**23**] and the argument of the proof of [**44**, Corollary 3.8 p. 89] imply:

THEOREM 1.1. *The morphism $BG \to BSing_\bullet^{\mathbb{A}^1}(G)$ induces an isomorphism*

$$H^*_{et}(BSing_\bullet^{\mathbb{A}^1}(G); M) \cong H^*_{et}(BG; M)$$

Using the previous results, Theorem 3 for G and F is now clearly a consequence of the following result, by taking for k the algebraic closure (in F) of the prime field:

THEOREM 1.2. *Assume k is algebraically closed. Let F be a separably closed field extension of k and let G be a semi-simple k-group. The natural ring homomorphism*

$$H^*_{et}(BSing^{\mathbb{A}^1}_\bullet(G);\mathbb{Z}/\ell) \to H^*(BSing^{\mathbb{A}^1}_\bullet(G)(F);\mathbb{Z}/\ell)$$

is an isomorphism.

This is our starting point. We will refer to this result as the "weak Friedlander conjecture". The space $BSing^{\mathbb{A}^1}_\bullet(G)$ is somehow much more flexible than the simplicial classifying space BG as it will appear in the next sections. For instance for G simply connected (as an algebraic k-group) $BSing^{\mathbb{A}^1}_\bullet(G)$ is a ($\mathbb{A}^1$-)simply connected space by Theorem 1.5 below, though in contrast BG has for only non trivial simplicial homotopy sheaf its π_1 which is G itself! The space $BSing^{\mathbb{A}^1}_\bullet(G)$ may be thought in some sense as the $+$-construction on BG. The morphism

$$BG \to BSing^{\mathbb{A}^1}_\bullet(G)$$

induces an isomorphism on $\mathbb{A}^1$-homology (see below) and $BSing^{\mathbb{A}^1}_\bullet(G)$ is $\mathbb{A}^1$-simply connected. Its homotopy sheaves in contrast to that of BG are unknown (expect the first one by the Hurewicz Theorem), but these are much more flexible objects!

1.2. The $\mathbb{A}^1$-local space $Sing^{\mathbb{A}^1}_\bullet(G)$ and its classifying space. In [**41**] we proved that $Sing^{\mathbb{A}^1}_\bullet(SL_n)$, for $n \geq 3$, is $\mathbb{A}^1$-local. Recall from [**44**] that a space $\mathcal{X}$ is $\mathbb{A}^1$-local if and only if the morphism to its $\mathbb{A}^1$-localization

$$\mathcal{X} \to L_{\mathbb{A}^1}(\mathcal{X})$$

is a simplicial weak equivalence. We proved in fact a much stronger property of $Sing^{\mathbb{A}^1}_\bullet(SL_n)$: it has the affine $\mathbb{A}^1$-Brown-Gersten property in the Nisnevich topology. This result has been generalized to any semi-simple k-group G by Wendt [**69**], except for the type SL_2, which has recently been settled by L.-F. Moser. Thus:

THEOREM 1.3. *([**41**] for SL_n, $n \geq 3$, Wendt [**69**] for type different of SL_2, Moser [**45**] for SL_n, $n \geq 2$) The Suslin-Voevodsky construction $Sing^{\mathbb{A}^1}_\bullet(G)$ of a split semi-simple k-group is $\mathbb{A}^1$-local.*

REMARK 1.4. 1) Let $X \in Sm_k$. We will say that X is *chain $\mathbb{A}^1$-local* if the Suslin-Voevodsky construction $Sing^{\mathbb{A}^1}_\bullet(X)$ on X is $\mathbb{A}^1$-local. In fact I do not know any example of X which is not $\mathbb{A}^1$-chain local.

2) Let M be a sheaf of abelian groups with transfers in the sense of Voevodsky [**67**]. Then one of the consequences of the results of Voevodsky on sheaves with transfers [**67**][**68**] is that the spaces $Sing^{\mathbb{A}^1}_\bullet(M)$, $BSing^{\mathbb{A}^1}_\bullet(M)$,

$K(Sing_\bullet^{\mathbb{A}^1}(M), n)$, etc ... are $\mathbb{A}^1$-local. The difficulty in proving Theorem 1.3 is of course that the sheaf of groups G is not abelian and has no reasonable structure of transfers. □

The following result is a non-trivial consequence of Theorem 1.3:

THEOREM 1.5. *1) The space $BSing_\bullet^{\mathbb{A}^1}(G)$ is $\mathbb{A}^1$-local.*

2) Assume furthermore that G is simply connected (and split semi-simple) as an algebraic k-group. Then the $\mathbb{A}^1$-local space $BSing_\bullet^{\mathbb{A}^1}(G)$ is $\mathbb{A}^1$-simply connected.

PROOF. The first statement follows from Theorem 1.3, [**41**, Thm 3.46] and the Theorem A.2. Now if G is split semi-simple simply connected, G is $\mathbb{A}^1$-connected by Theorem A.2 again, and thus BG is simply $\mathbb{A}^1$-connected. □

REMARK 1.6. This is the only place where we use the assumption *char* $(k) \neq 2$. As so far Theorem 1.3 only holds in characteristic not 2. □

2. The Friedlander conjecture and the rigidity property

2.1. The rigidity properties at ℓ. We now introduce the rigidity property at ℓ concerning a space. A discussion of rigidity properties and various implications towards the Friedlander-Milnor conjecture may be found in [**30**, §5.3]. We consider 3 notions, inspired by the rigidity Theorem of [**63**] and the approach of [**28**], though in fact we are able to prove the strongest possible one.

Below we show that if the space $BSing_\bullet^{\mathbb{A}^1}(G)$ has the rigidity property then the weak Friedlander conjecture holds for G and any separably closed field extension F of k.

DEFINITION 2.1. Let $\mathcal{X}$ be a space.

1) We will say that $\mathcal{X}$ has the *weak rigidity property* at ℓ if for any extension $k \subset F$ with F a separably closed field, the natural map of simplicial sets $\mathcal{X}(k_s) \to \mathcal{X}(F)$ induces an isomorphism in mod ℓ homology:
$$H_*(\mathcal{X}(k_s); \mathbb{Z}/\ell) \cong H_*(\mathcal{X}(F); \mathbb{Z}/\ell)$$
(Here $k_s \subset F$ means the separable closure of k in F, which is of course separably closed itself.) When no confusion can occur on ℓ which is fixed, we will just say "$\mathcal{X}$ has the weak rigidity property" instead of "$\mathcal{X}$ has the weak rigidity property at ℓ".
2) We will say that $\mathcal{X}$ has the *rigidity property* at ℓ if for each strict henselization $\mathcal{O}_{X,x}^{sh}$ of the local ring of a point $x \in X \in Sm_k$ with separably closed residue field $\kappa(x)_s$, the natural map of simplicial sets $\mathcal{X}(\mathcal{O}_{X,x}^{sh}) \to \mathcal{X}(\kappa(x)_s)$ induced by restriction to the closed point induces an isomorphism in mod ℓ homology:
$$H_*(\mathcal{X}(\mathcal{O}_{X,x}^{sh}); \mathbb{Z}/\ell) \cong H_*(\mathcal{X}(\kappa(x)_s); \mathbb{Z}/\ell)$$

When no confusion can occur on ℓ, we will just say "$\mathcal{X}$ has the rigidity property" instead of "$\mathcal{X}$ has the rigidity property at ℓ".

3) We will say that $\mathcal{X}$ has the *strong rigidity property* at ℓ if for each henselization $\mathcal{O}^h_{X,x}$ of the local ring of a point $x \in X \in Sm_k$ with residue field $\kappa(x)$, the natural map of simplicial sets $\mathcal{X}(\mathcal{O}^h_{X,x}) \to \mathcal{X}(\kappa(x))$ induced by restriction to the closed point induces an isomorphism in mod ℓ homology:

$$H_*(\mathcal{X}(\mathcal{O}^h_{X,x}); \mathbb{Z}/\ell) \cong H_*(\mathcal{X}(\kappa(x)); \mathbb{Z}/\ell)$$

When no confusion can occur on ℓ, we will just say "$\mathcal{X}$ has the strong rigidity property" instead of "$\mathcal{X}$ has the strong rigidity property at ℓ".

EXAMPLE 2.2. The simplicial classifying space $B\mathbb{G}_m$ of the multiplicative group $\mathbb{G}_m$ has the strong rigidity property at ℓ. Indeed to check this we have to prove that for each henselization $A := \mathcal{O}^h_{X,x}$ of a point $x \in X \in Sm_k$ with residue field $\kappa = \kappa(x)$ the morphism "restriction to the closed point"

$$B(A^\times) \to B(\kappa^\times)$$

induces an isomorphism on mod ℓ-homology for ℓ different from the characteristic of κ. It suffices to check that the kernel K of the surjective group homomorphism $A^\times \to \kappa^\times$ has trivial mod ℓ-homology; for this it suffices to observe that K is uniquely ℓ-divisible, which follows directly from the classical definition of an henselian local ring.

Observe that $\mathbb{G}_m$ being $\mathbb{A}^1$-rigid, $B\mathbb{G}_m = BSing_\bullet^{\mathbb{A}^1}(\mathbb{G}_m)$.

Observe however that the multiplicative group $\mathbb{G}_m$ itself, considered as a space (placed in simplicial degree 0) has not the rigidity property at ℓ. □

REMARK 2.3. 1) The main result of this work is to prove that more generally, the space $BSing_\bullet^{\mathbb{A}^1}(G)$ has the strong rigidity property at ℓ, for G split semi-simple (see Theorem 2.17).

2) It doesn't follow that the space BG itself has the rigidity property at ℓ. It sounds reasonable to believe this anyway. This would follow from our main result (Theorem 2.17) and from the following property of "homotopy invariance for group cohomology on henselian rings": let $\mathcal{O}^h_{X,x}$ be the Henselization of a point $x \in X \in Sm_k$ then the morphism

$$H_*(BG(\mathcal{O}^h_{X,x}); \mathbb{Z}/\ell) \to H_*(BG(\mathcal{O}^h_{X,x}[T_1, \ldots, T_n]); \mathbb{Z}/\ell)$$

is an isomorphism for each n. □

The following Lemma will be useful:

LEMMA 2.4. *Assume that $\mathcal{F} \to \mathcal{E} \to \mathcal{B}$ is a simplicial fibration between fibrant pointed spaces, with $\mathcal{B}$ connected. Assume that for any $x \in X \in Sm_k$ the action of $\pi_1(\mathcal{B}(\mathcal{O}^h_{X,x}))$ on the mod ℓ-homology of $\mathcal{F}((\mathcal{O}^h_{X,x}))$ is trivial (for instance if $\mathcal{B}$ is simply connected or if the fibration is up to weak equivalence* *a principal fibration with group $\mathcal{G} \cong \mathcal{F}$ connected). Then if two of $\mathcal{B}$, $\mathcal{E}$ or $\mathcal{F}$ have the rigidity property of type $i \in \{1,2,3\}$, then so has the third.*

PROOF. Indeed, let us do the case of type 3, the strong rigidity property. For each point $x \in X \in Sm_k$ with residue field $\kappa(x)$ we may consider the two Serre homology spectral sequences with mod ℓ coefficients of the two fibrations sequences of simplicial sets $\mathcal{F}(\mathcal{O}^h_{X,x}) \to \mathcal{E}(\mathcal{O}^h_{X,x}) \to \mathcal{B}(\mathcal{O}^h_{X,x})$ and $\mathcal{F}(\kappa(x)) \to \mathcal{E}(\kappa(x)) \to \mathcal{B}(\kappa(x))$. From the assumptions, both E^2 terms are of the form $H_*(\mathcal{B};\mathbb{Z}/\ell) \otimes H_*(\mathcal{F};\mathbb{Z}/\ell)$ (meaning that the local coefficients systems are constant). Using the naturality of the Serre spectral sequence, the rest of the argument is classical. □

REMARK 2.5. For instance, assume that $\mathcal{F} \to \mathcal{E} \to \mathcal{B}$ is a simplicial fibration between fibrant pointed spaces with $\mathcal{B}$ simply connected and $\mathcal{E}$ weakly contractible. In other words $\mathcal{F}$ has the weak simplicial homotopy type of the simplicial loop space $\Omega(\mathcal{B})$. Then the previous Lemma tells us that the following properties are equivalent:

(1) $\mathcal{B}$ has the rigidity property of type i,
(2) $\Omega(\mathcal{B})$ has the rigidity property of type i.

Example 2.2 shows that in general the result (and thus Lemma 2.4) doesn't hold without assumptions on the π_1 of $\mathcal{B}$. □

Let us mention an application without proof:

LEMMA 2.6. *For any split semi-simple k-group G with universal covering $\tilde{G} \to G$ (in the sense of algebraic group theory), if $BSing_{\bullet}^{\mathbb{A}^1}(\tilde{G})$ has the rigidity property, then $BSing_{\bullet}^{\mathbb{A}^1}(G)$ has the rigidity property.*

In what follows we give a not very surprising characterization of pointed simply connected space $\mathcal{X}$ which have the rigidity property in term of their simplicial homotopy sheaves. We leave to the reader to state and prove the analogue statement involving the weak (resp. strong) rigidity property. Recall that for M an abelian group we let respectively ${}_{[\ell]}M$ and M/ℓ be the kernel and the cokernel of the homomorphism "multiplication by ℓ": $M \xrightarrow{\ell} M$.

THEOREM 2.7. *Let $\mathcal{X}$ be a pointed simply connected space. The following conditions are equivalent:*

(i) *$\mathcal{X}$ has the rigidity property;*
(ii) *for each integer $n \geq 2$ and for each henselization $\mathcal{O}^h_{X,x}$ of the local ring of a point $x \in X \in Sm_k$ with residue field $\kappa(x)$, the kernel of the natural epimorphism $\pi_n(\mathcal{X})(\mathcal{O}^h_{X,x}) \to \pi_n(\mathcal{X})(\kappa(x))$ is uniquely ℓ-divisible.*
(iii) *for each integer $n \geq 2$ and for each henselization $\mathcal{O}^h_{X,x}$ of the local ring of a point $x \in X \in Sm_k$ with residue field $\kappa(x)$, the two homorphisms ${}_{[\ell]}\pi_n(\mathcal{X})(\mathcal{O}^h_{X,x}) \to {}_{[\ell]}\pi_n(\mathcal{X})(\kappa(x))$ and $\pi_n(\mathcal{X})(\mathcal{O}^h_{X,x})/\ell \to \pi_n(\mathcal{X})(\kappa(x))/\ell$ are isomorphisms.*

PROOF. The equivalence $(ii) \Leftrightarrow (iii)$ is clear. The implication $(ii) \Rightarrow (i)$ is proven using the Postnikov tower $\{P^{(n)}(\mathcal{X})\}_n$ of $\mathcal{X}$ and follows by induction on n from the fact that for each n the map $K(\pi_n(\mathcal{X})(\mathcal{O}^h_{X,x}), n) \to K(\pi_n(\mathcal{X})(\kappa(x)), n)$ induces by assumption an isomorphism on mod ℓ homology. To prove the converse implication $(i) \Rightarrow (ii)$ one uses induction on r and the Lemma 2.8 below. □

LEMMA 2.8. *Let $f : K \to L$ be a morphism of pointed simply connected simplicial sets which admits a section in the pointed homotopy category, let ℓ be a prime number and let $r \geq 1$ be a natural number. The following conditions are equivalent:*

(i) *the morphism f induces an isomorphism on $H_*(-;\mathbb{Z}/\ell)$ for $* \leq r$;*
(ii) *let Γ be the homotopy fiber of f; then the homology $H_*(\Gamma;\mathbb{Z}/\ell)$ is trivial for $* \leq r$;*
(iii) *the kernel of the (split) epimorphism $\pi_i(K) \to \pi_i(L)$ is uniquely ℓ-divisible for $i \leq r-1$ and ℓ-divisible for $i = r$. In other words $_{[\ell]}\pi_i(K) \to {}_{[\ell]}\pi_i(L)$ is an isomorphism for $i \leq r-1$ and $\pi_i(K)/\ell \to \pi_i(L)/\ell$ is an isomorphism for $i \leq r$.*

2.2. Reformulation of the rigidity property in term of simplicial chain complex. We let $\mathcal{A}b(k)$ denote the abelian category of sheaves of abelian groups in the Nisnevich topology on Sm_k, and $\mathcal{C}_*(\mathcal{A}b(k))$ the category of chain complexes in $\mathcal{A}b(k)$ (with differential of degree -1). Given a space $\mathcal{X}$ we let $\mathbb{Z}(\mathcal{X})$ denote the free simplicial sheaf of abelian groups generated by $\mathcal{X}$, that is to say the sheafification of the presheaf $U \mapsto \mathbb{Z}(\mathcal{X}(U))$ obtained by taking (degreewise) the free abelian group generated by the simplicial set $\mathcal{X}(U)$. This is a simplicial sheaf of abelian groups, and we may associate to it its normalized chain complex $C_*(\mathcal{X};\mathbb{Z}) \in \mathcal{C}_*(\mathcal{A}b(k))$ (see for instance [**9**] for the definition of normalized chain complex). This chain complex is called the *simplicial chain complex*[3] of $\mathcal{X}$. We let $C_*(\mathcal{X};\mathbb{Z}/\ell)$ be the mod ℓ version, which means the tensor product $C_*(\mathcal{X};\mathbb{Z}) \otimes \mathbb{Z}/\ell = C_*(\mathcal{X};\mathbb{Z})/\ell$.

The n-th homology sheaf of $C_*(\mathcal{X};\mathbb{Z})$ (resp. $C_*(\mathcal{X} : \mathbb{Z}/\ell)$) is called the n-th simplicial homology sheaf of $\mathcal{X}$ with $\mathbb{Z}$ (resp. $\mathbb{Z}/\ell$) coefficients and is denoted by $H_n(\mathcal{X};\mathbb{Z})$ (resp. $H_n(\mathcal{X};\mathbb{Z}/\ell)$).

DEFINITION 2.9. 1) A chain complex C_* of $\mathbb{Z}/\ell$-modules will be said to have the rigidity property (at ℓ) if and only if for each henselization $\mathcal{O}^h_{X,x}$ of the local ring of a point $x \in X \in Sm_k$ with residue field $\kappa(x)$, the natural morphism of chain complexes of $\mathbb{Z}/\ell$-modules

$$C_*(\mathcal{O}^h_{X,x}) \to C_*(\kappa(x))$$

is a quasi-isomorphism.

[3]as opposed to the $\mathbb{A}^1$-chain complex which will appear later

2) A chain complex C_* of abelian groups will be said to have the rigidity property (at ℓ) if and only if for each henselization $\mathcal{O}^h_{X,x}$ of the local ring of a point $x \in X \in Sm_k$ with residue field $\kappa(x)$, the natural morphism of chain complexes of $\mathbb{Z}/\ell$-modules

$$C_*(\mathcal{O}^h_{X,x}) \otimes^L \mathbb{Z}/\ell \to C_*(\kappa(x)) \otimes^L \mathbb{Z}/\ell$$

is a quasi-isomorphism (in other words if $C_* \otimes^L \mathbb{Z}/\ell$ has the rigidity property).

EXAMPLE 2.10. 1) A complex of sheaves with transfers in the sense of Voevodsky [**67**] whith $\mathbb{A}^1$-invariant homology sheaves has the rigidity property at ℓ by [**63**].

2) Let M be a sheaf of $\mathbb{Z}/\ell$-modules. Then the complex M (placed in degree 0) has the rigidity property at ℓ if and only if for each henselization $\mathcal{O}^h_{X,x}$ of the local ring of a point $x \in X \in Sm_k$ with residue field $\kappa(x)$, the natural epimorphism of $\mathbb{Z}/\ell$-modules $M(\mathcal{O}^h_{X,x}) \to M(\kappa(x)$ is an isomorphism.

3) Let M be a sheaf of abelian groups. Then the complex M (placed in degree 0) has the rigidity property at ℓ if and only if the sheaves M/ℓ and ${}_{[\ell]}M$ (kernel of the multiplication by ℓ) have the rigidity property. Of course as these two sheaves are sheaves of $\mathbb{Z}/\ell$-modules, we may use the characterization given in the preceding point. □

LEMMA 2.11. *1) Let $\mathcal{X}$ be a space. The following conditions are equivalent:*

(i) *the space $\mathcal{X}$ has the rigidity property at ℓ;*
(ii) *the chain complex $C_*(\mathcal{X}; \mathbb{Z}/\ell)$ has the rigidity property at ℓ.*

2) Let C_* be a chain complex of sheaves of $\mathbb{Z}/\ell$-modules and let KC_* be its corresponding Eilenberg-MacLane space. If $H_*(C_*) = 0$ for $* < 0$, the following conditions are equivalent:

(i) *the space KC_* has the rigidity property at ℓ;*
(ii) *the chain complex C_* has the rigidity property.*

3) Let C_* be a chain complex of sheaves of abelian groups and let KC_* be its corresponding Eilenberg-MacLane space. If $H_*(C_*) = 0$ for $* \leq 0$, the following conditions are equivalent:

(i) *the space KC_* has the rigidity property at ℓ;*
(ii) *the chain complex C_* has the rigidity property.*

PROOF. The point 1) is clear. The point 2) follows from the following classical fact: a morphism $A_\bullet \to B_\bullet$ between two simplicial $\mathbb{Z}/\ell$-modules is a simplicial weak equivalence if and only if it induces a isomorphism in the mod ℓ homology of the underlying simplicial sets. The third point follows easily from this, taking into account the connectivity assumption. □

REMARK 2.12. Observe that one can't remove the assumption $H_*(C_*) = 0$ for $* \leq 0$ in the last point, as for instance the chain complex $\mathbb{G}_m$ itself has the rigidity property at ℓ (compute $H_*(\mathbb{G}_m \otimes^L \mathbb{Z}/\ell)$ and check it!) but the space $\mathbb{G}_m$ doesn't have the rigidity property at ℓ.

We may now prove:

THEOREM 2.13. *Assume k is algebraically closed and let $\mathcal{X}$ be a space which has the rigidity property at ℓ. Then the canonical morphism of complexes of sheaves*

$$C_*(\mathcal{X}(k);\mathbb{Z}/\ell) \to C_*(\mathcal{X};\mathbb{Z}/\ell)$$

is a quasi-isomorphism in the étale topology (where the left hand side means the associated complex of constant sheaves). Equivalently, for any $n \in \mathbb{N}$, the sheaf $H_n(\mathcal{X};\mathbb{Z}/\ell)$ is constant in the étale topology.

Finally the natural ring homomorphism (see below for the precise definition):

$$H^*_{et}(\mathcal{X};\mathbb{Z}/\ell) \to H^*(\mathcal{X}(k);\mathbb{Z}/\ell) \tag{2.1}$$

is an isomorphism.

PROOF. The first statement follows from the definition of the rigidity property at ℓ, Lemma 2.11 above and Lemma 2.15 below, just by using the property of Definition 2.9 for henselization of closed points of smooth k-schemes.

The last assertion follows from this and the fact that if we let $\mathbb{Z}/\ell_{et}(k)$ denote the abelian category of sheaves of $\mathbb{Z}/\ell$-modules [**23**], and $D(\mathbb{Z}/\ell_{et}(k))$ its derived category, then for any space $\mathcal{X}$, by the very definition of étale cohomology, one has a canonical isomorphism

$$Hom_{D(\mathbb{Z}/\ell_{et}(k))}(C_*(\mathcal{X};\mathbb{Z}/\ell)_{et};\mathbb{Z}/\ell[*]) \cong H^*_{et}(\mathcal{X};\mathbb{Z}/\ell)$$

where $C_*(\mathcal{X};\mathbb{Z}/\ell)_{et}$ is the simplicial mod ℓ chain complex of $\mathcal{X}$ in the étale topology, in other words the sheafification of $C_*(\mathcal{X};\mathbb{Z}/\ell)$ in the étale topology. The canonical (iso)morphism mentioned in the statement of the Theorem is thus the composition:

$$\begin{aligned} H^*_{et}(\mathcal{X};\mathbb{Z}/\ell) &= Hom_{D(\mathbb{Z}/\ell_{et}(k))}(C_*(\mathcal{X};\mathbb{Z}/\ell)_{et};\mathbb{Z}/\ell[*]) \\ &\cong Hom_{D(\mathbb{Z}/\ell_{et}(k))}(C_*(\mathcal{X}(k);\mathbb{Z}/\ell)_{et};\mathbb{Z}/\ell[*]) \\ &= H^*_{et}(\mathcal{X}(k);\mathbb{Z}/\ell) = H^*(\mathcal{X}(k);\mathbb{Z}/\ell) \end{aligned}$$

induced by the (étale) quasi-isomorphism (2.1). The last identification $H^*_{et}(\mathcal{X}(k);\mathbb{Z}/\ell) = H^*(\mathcal{X}(k);\mathbb{Z}/\ell)$ follows from the fact that any constant sheaf of $\mathbb{Z}/\ell$-modules in the étale topology is projective (as k is algebraically closed). ☐

REMARK 2.14. There is no obvious sheaf theoretic reformulation of the weak or strong rigidity property. For instance observe that the sheaf $\underline{\mathbf{K}}^M_n$, $n \geq 1$ say, has the strong rigidity property. In particular the associated sheaf to $\underline{\mathbf{K}}^M_n/\ell$ in the étale topology is indeed trivial (easy to check, any unit is an ℓ-power locally in the étale toplogy). However the strong rigidity property tells us much more: for any henselization $\mathcal{O}^h_{X,x}$ of a point x in a smooth k-scheme X the morphism

$$\underline{\mathbf{K}}_n^M/\ell(\mathcal{O}^h_{X,x}) \to \underline{\mathbf{K}}_n^M/\ell(\kappa(x))$$

is an isomorphism.

We will only care on the rigidity property in this work as it is sufficient to prove Theorem 3. □

LEMMA 2.15. *Let M be a sheaf of abelian group in the étale topology on Sm_k, k algebraically closed. Assume that for each henselization $\mathcal{O}^h_{X,x}$ of the local ring of a closed point $x \in X \in Sm_k$ (with residue field automatically k), the morphism $M(\mathcal{O}^h_{X,x}) \to M(k)$ is an isomorphism. Then M is a constant sheaf in the étale topology, that is to say the morphism $M(k)_{et} \to M$ is an isomorphism.*

PROOF. The morphism $M(k)_{et} \to M$ is an injective morphism because for X irreducible smooth over k, the morphism $M(k)_{et} \to M(X)$ can be identified with the morphism $M(k) \to M(X)$ induced by the structural morphism $X \to Spec(k)$. Thus we may replace M by the cokernel of this morphism and assume that $M(k) = 0$. We thus have to prove that $M = 0$. Take $\alpha \in M(X)$. For each closed point $x \in X(k)$, $\alpha_x = 0 \in M(\mathcal{O}^h_{X,x})$ (because $M(k) = 0$). This means that there is an étale morphism $V_x \to X$ whose image contains x such that $\alpha|V_x = 0$. But the family $\{V_x \to X\}_x$ is a covering family of X in the étale topology, so that $\alpha = 0$. □

REMARK 2.16. 1) Let $\mathcal{X}$ be a pointed simply connected space. Let $\mathcal{X} \to \mathcal{X}[\frac{1}{\ell}]$ be the universal morphism obtained by inverting ℓ in the homotopy sheaves of $\mathcal{X}$ (this is constructed as in [**58**]). The homotopy fiber $\mathcal{X}_{\ell-tor}$ of the morphism $\mathcal{X} \to \mathcal{X}[\frac{1}{\ell}]$ is a space whose homotopy sheaves are ℓ-torsion. If $\mathcal{X}$ satisfies has the rigidity property at ℓ, it is easy to check that $\mathcal{X}_{\ell-tor}$ also has the rigidity property at ℓ. Moreover if k is algebraically closed, then $\mathcal{X}_{\ell-tor}(k) \to \mathcal{X}$ is a simplicial weak equivalence in the étale topology. In other words, $\mathcal{X}_{\ell-tor}$ becomes constant in the étale topology.

2) One may deduce from the Theorem that if a space $\mathcal{X}$ has the rigidity property at ℓ, the obvious morphism $\mathcal{X}(k) \to \mathcal{X}$ induces an equivalence of pro-ℓ spaces $\mathcal{X}(k)^{\wedge}_{\ell} \to \hat{\mathcal{X}}_{\ell,et}$ after taking the étale homotopy type functor [**3, 13**], where the left hand side means the pro-ℓ-completion of the simplicial set $\mathcal{X}(k)$. □

Taking Theorem 2.13 into account, we have reduced the proof of Theorem 1.2 and thus of Theorem 3 to the following:

THEOREM 2.17. *For any perfect field k and any split semi-simple k-group G, the space*

$$BSing_\bullet^{\mathbb{A}^1}(G)$$

has the rigidity property.

REMARK 2.18. As we already mentioned, in fact, we may prove the strong rigidity property. See [**42**][**43**]. □

Now to start proving Theorem 2.17, we need at last to let the $\mathbb{A}^1$-chain complex come in the game.

3. The rigidity property and the $\mathbb{A}^1$-chain complex

3.1. $\mathbb{A}^1$-chain complexes and the $\mathbb{A}^1$-lower central series. We want now to connect the rigidity property of a space $\mathcal{X}$, or equivalently (see the previous section) of its simplicial chain complex $C_*(\mathcal{X})$, to the rigidity property of its $\mathbb{A}^1$-chain complex $C_*^{\mathbb{A}^1}(\mathcal{X})$. The latter is still a mysterious object, obtained from $C_*(\mathcal{X})$ by applying a rather abstract functor $L_{\mathbb{A}^1}^{ab}$ of $\mathbb{A}^1$-localization of chain complexes. But on the other hand it appears also to be more flexible. First it is clear that the morphism

$$C_*^{\mathbb{A}^1}(BG) \to C_*^{\mathbb{A}^1}(BSing_\bullet^{\mathbb{A}^1}(G))$$

is a quasi-isomorphism. At this point there is no difference between the two spaces involved. One of the crucial steps in our approach is to be able to observe that $C_*^{\mathbb{A}^1}(G)$, and then $C_*^{\mathbb{A}^1}(BSing_\bullet^{\mathbb{A}^1}(G))$, lies in the thick subcategory of chain complexes generated by the $C_*^{\mathbb{A}^1}((\mathbb{G}_m)^{\wedge n})$'s, as it follows from the Bruhat decomposition. This wouldn't be possible if we would stay at the level of simplicial chain complexes.

We start by giving some recollections on $\mathbb{A}^1$-chain complexes from [**38, 41**]. We let $D(\mathcal{A}b(k))$ be the derived category of $\mathcal{A}b(k)$, that is to say the category obtained from $\mathcal{C}_*(\mathcal{A}b(k))$ by inverting the quasi-isomorphisms. This category is the homotopy category of a model category structure [**49**] on $\mathcal{C}_*(\mathcal{A}b(k))$ with weak equivalences the quasi-isomorphisms and cofibrations the monomorphisms[4]. In the case of the category of abelian sheaves in the Nisnevich topology, which has a set of generators of finite cohomological dimension, this is well known, see [**41**, Section 3.2].

We let $D^{\mathbb{A}^1-loc}(\mathcal{A}b(k)) \subset D(\mathcal{A}b(k))$ the full subcategory consisting of $\mathbb{A}^1$-local chain complexes as defined in [**41**, Section 3.2 p. 94]. From *loc. cit.* this inclusion admits a left adjoint called the $\mathbb{A}^1$-localization functor

$$L_{\mathbb{A}^1}^{ab} : D(\mathcal{A}b(k)) \subset D_{\mathbb{A}^1}(\mathcal{A}b(k))$$

which induces an equivalence of categories $D_{\mathbb{A}^1}(\mathcal{A}b(k)) \cong D^{\mathbb{A}^1-loc}(\mathcal{A}b(k))$, where $D_{\mathbb{A}^1}(\mathcal{A}b(k))$ is the $\mathbb{A}^1$-derived category.

The main result of [**38**] is that the functor $L_{\mathbb{A}^1}^{ab}$ preserves connectivity of chain complexes. As a consequence, we proved in *loc. cit.* that if $\mathcal{A}b_{\mathbb{A}^1}(k) \subset \mathcal{A}b(k)$ denote the full subcategory of strictly $\mathbb{A}^1$-invariant sheaves, a chain

[4]The existence of this model category structure on the category of complexes in a Grothendieck abelian category is due independently to several authors, amongst them A. Joyal, and the author, but so far unpublished

complex C_* is $\mathbb{A}^1$-local if and only if its homology sheaves $H_*(C_*)$ are in $\mathcal{A}b_{\mathbb{A}^1}(k)$.

REMARK 3.1. If $C_* \in \mathcal{C}_*(\mathcal{A}b(k))$ has trivial homology sheaves in degrees ≤ 0, it is proven[5] in [**41**, Corollary 3.24 p. 100] that the morphism of spaces induced by the $\mathbb{A}^1$-localization functor $KC_* \to KL^{ab}_{\mathbb{A}^1}C_*$ is an $\mathbb{A}^1$-weak equivalence to an $\mathbb{A}^1$-local space. Thus in that case the canonical morphism

$$L_{\mathbb{A}^1}KC_* \to KL^{ab}_{\mathbb{A}^1}C_*$$

is a simplicial weak equivalence.

However in general (if $H_0C_* \neq 0$) this is not true, and for this reason we used different notations for the $\mathbb{A}^1$-localization functor for spaces and for complexes. However in each of the situations we will meet the previous result will imply that we may identify those using the previous simplicial weak-equivalence. We will use frequently this identification. □

Let $\mathcal{X}$ be a space and let $C_*(\mathcal{X};\mathbb{Z})$ be its normalized chain complex, see Section 2.2. The $\mathbb{A}^1$-localization $L^{ab}_{\mathbb{A}^1}(C_*(\mathcal{X};\mathbb{Z}))$ is denoted by $C^{\mathbb{A}^1}_*(\mathcal{X};\mathbb{Z})$, or sometimes simply $C^{\mathbb{A}^1}_*(\mathcal{X})$, and is called the $\mathbb{A}^1$-*chain complex* of $\mathcal{X}$.

From the previous results, it follows that the homology sheaves $H^{\mathbb{A}^1}_i$ $(\mathcal{X};\mathbb{Z})$ of $C^{\mathbb{A}^1}_*(\mathcal{X};\mathbb{Z})$ are strictly $\mathbb{A}^1$-invariant sheaves and vanish for $i < 0$. This are called the $\mathbb{A}^1$-homology sheaves of the space $\mathcal{X}$.

REMARK 3.2. Even if $\mathcal{X}$ is an $\mathbb{A}^1$-local space, the chain complex $C_*(\mathcal{X};\mathbb{Z})$ has no reason to be $\mathbb{A}^1$-local! See Remark 3.11 below. □

When $\mathcal{X}$ is pointed, one may consider the reduced $\mathbb{A}^1$-chain complex $\tilde{C}^{\mathbb{A}^1}_*(\mathcal{X};\mathbb{Z})$, as well the reduced $\mathbb{A}^1$-homology sheaves $\tilde{H}^{\mathbb{A}^1}_*(\mathcal{X};\mathbb{Z})$, by removing the factor $\mathbb{Z}$ in $C^{\mathbb{A}^1}_*(\mathcal{X};\mathbb{Z})$ corresponding to the base point. We may also consider as in the simplicial case the $\mathbb{A}^1$-chain complexes and homology sheaves with $\mathbb{Z}/\ell$-coefficients: $C^{\mathbb{A}^1}_*(\mathcal{X};\mathbb{Z}/\ell)$ and $H^{\mathbb{A}^1}_*(\mathcal{X};\mathbb{Z}/\ell)$.

Now we introduce one of the main new technical tool: the $\mathbb{A}^1$-lower central series. Let us start by recalling the classical one.

The lower central series for sheaves of groups. We assume the reader familiar with the usual lower central series for groups, which can be found for instance in [**9**]. Let G be a sheaf of groups (on $(Sm_k)_{Nis}$). Its "naive" lower central series $\dots\Gamma_rG \subset \Gamma_{r-1}G \subset \Gamma_1G = G$ is defined by sheafification of the usual one on G. It is the smallest descending filtration $\dots \subset \Gamma_{r+1} \subset \Gamma_r \subset \dots \subset G$ of G by normal subgroups such that $[G,\Gamma_r] \subset \Gamma_{r+1}$.

It is well known (*cf* loc. cit.) that there is a structure of graded Lie algebra on the associated graded (sheaf of) abelian group(s) $Gr_*(G) = \oplus\Gamma_r(G)/\Gamma_{r+1}(G)$, that $Gr_1(G) = G_{ab}$ the abelianization of G, and that the induced morphism

$$L_*(G_{ab}) \twoheadrightarrow Gr_*(G)$$

[5]This is at this place that we need for the moment that $char(k) \neq 2$.

from the free graded Lie algebra $L_*(G_{ab})$ is an epimorphism, and an isomorphism for G stalkwise free, see Theorem 7.5 of [**9**].

The Kan construction $G(\mathcal{X})$. Let $\mathcal{X}$ be a reduced pointed space, meaning that $\mathcal{X}_0$ is equal to the point. One denotes by $G(\mathcal{X})$ the sheafification of the Kan G construction [**9**, Definition 3.15 p. 133] on the presheaf $U \mapsto G(\mathcal{X}(U))$. It is a simplicial sheaf of groups, stalkwise free.

Standard homotopical algebra provides a canonical isomorphism in the pointed simplicial homotopy category of spaces between $R\Omega(\mathcal{X})$ and $G(\mathcal{X})$.

Using the main results of [**8**] we get:

THEOREM 3.3. *Let $\mathcal{X}$ be a simplicially* 1*-connected reduced pointed space. Then for each $r \geq 1$ the morphism*

$$G(\mathcal{X}) \to G(\mathcal{X})/\Gamma_r(G(\mathcal{X}))$$

is at least $\{1 + log_2(r)\}$ simplicially connected.

The fiber of the morphism $G(\mathcal{X})/\Gamma_r(G(\mathcal{X})) \to G(\mathcal{X})/\Gamma_{r-1}(G(\mathcal{X}))$ is of course isomorphism as a simplicial sheaf of groups to $\Gamma_{r-1}(G(\mathcal{X}))/\Gamma_r(G(\mathcal{X}))$. From what we recall above, $G(\mathcal{X})$ being stalkwise free, this simplicial abelian group is isomorphic to the simplicial sheaf of abelian groups

$$F_r := L_{r-1}(G(\mathcal{X})_{ab})$$

obtained by degreewise applying the free Lie algebra functor in degree $r-1$ to the simplicial abelian group $G(\mathcal{X})_{ab}$. This simplicial sheaf of abelian groups F_r is also (approximatively) $\{1 + log_2(r)\}$ simplicially connected.

The $\mathbb{A}^1$-lower central series. We may now deduce from the unstable $\mathbb{A}^1$-connectivity result of [**41**] the following:

THEOREM 3.4. *Let $\mathcal{X}$ be a* 1*-connected $\mathbb{A}^1$-local reduced pointed space. Then for each $r \geq 1$ the morphism*

$$G(\mathcal{X}) \cong L_{\mathbb{A}^1}(G(\mathcal{X})) \to L_{\mathbb{A}^1}(G(\mathcal{X})/\Gamma_r(G(\mathcal{X})))$$

is $\{1 + log_2(r)\}$-connected.

PROOF. From [**41**] the exact sequence of simplicial sheaf of groups $\Gamma_r(G(\mathcal{X})) \to G(\mathcal{X}) \to G(\mathcal{X})/\Gamma_r(G(\mathcal{X}))$ is automatically an $\mathbb{A}^1$-fibration sequence. Now from *loc. cit.* the space $\Gamma_r(G(\mathcal{X}))$ being simplicially $\{1 + log_2(r)\}$ connected, is $\{1 + log_2(r)\}$ $\mathbb{A}^1$-connected. □

For a reduced Kan simplicial set K recall [**9**, p. 125] that its "path complex" $\mathcal{P}K$ is the simplicial set with $\mathcal{P}K_n := K_{n+1}$ and $d_i := d_i + 1$, $s_i := s_{i+1}$. The morphism d_0 induces a Kan fibration $\mathcal{P}K \to K$ with fiber ΩK the loop space of K. This construction generalizes to reduced simplicial sheaf of sets.

If $\mathcal{X}$ is a pointed space, we let $\tilde{\mathbb{Z}}(\mathcal{X})$ be the free sheaf of abelian groups generated by $\mathcal{X}$ modulo the relation $* = 0$ (base point equals 0). In other words we have a canonical direct sum decomposition in the category of simplicial sheaves of abelian groups of the form:

$$\mathbb{Z}(\mathcal{X}) = \mathbb{Z} \oplus \tilde{\mathbb{Z}}(\mathcal{X})$$

The associated normalized chain complex to the simplicial sheaf of abelian groups $\mathbb{Z}(\mathcal{X})$ is called the reduced chain complex of $\mathcal{X}$ and is denoted by $\tilde{C}_*(\mathcal{X})$ of $\mathcal{X}$. The Dold-Kan correspondance gives an isomorphism of simplicial sheaves of abelian groups (see [**9**])

$$\tilde{\mathbb{Z}}(\mathcal{X}) \cong K(\tilde{C}_*(\mathcal{X}))$$

An easy inspection in the previous definitions gives the following standard fact:

LEMMA 3.5. *There exists canonical isomorphisms of simplicial sheaves of abelian groups*

$$G(\mathcal{X})_{ab} \cong \Omega(\tilde{\mathbb{Z}}(\mathcal{X})) \cong \Omega K(\tilde{C}_*(\mathcal{X})) \cong K(\tilde{C}_*(\mathcal{X})[-1])$$

It follows that the space F_s considered above is in fact (up to canonical isomorphism) $F_s = L_{s-1}(K(\tilde{C}_*(\mathcal{X})[-1]))$, a functor of the reduced simplicial chain complex $\tilde{C}_*(\mathcal{X})$. Let us denote by $F_s^{\mathbb{A}^1}$ the $\mathbb{A}^1$-localization of F_s. We will use the simplicial fibration sequences

$$F_s^{\mathbb{A}^1} \to L_{\mathbb{A}^1}(G(\mathcal{X})/\Gamma_s(G(\mathcal{X}))) \to L_{\mathbb{A}^1}(G(\mathcal{X})/\Gamma_{s-1}(G(\mathcal{X}))) \tag{3.1}$$

It is (up to simplicial weak equivalence) a principal fibration sequence with connected structural group

$$F_s^{\mathbb{A}^1} \cong L_{\mathbb{A}^1}(L_{s-1}(K(\tilde{C}_*(\mathcal{X})[-1]))) \tag{3.2}$$

An $\mathbb{A}^1$-quasi-isomorphism $C_* \to D_*$ is a morphism of chain complexes which induces an isomorphism in the $\mathbb{A}^1$-derived category $D_{\mathbb{A}^1}(\mathcal{A}b(k))$ or, equivalently, which induces a quasi-isomorphism after applying the $\mathbb{A}^1$-localization functor $L^{ab}_{\mathbb{A}^1}$. For instance $C_* \to L^{ab}_{\mathbb{A}^1}(C_*)$ is an $\mathbb{A}^1$-quasi isomorphism. The following result may be proven along the "classical" lines (see [**11**]) for instance.

THEOREM 3.6. *Let $s \geq 1$ be an integer and let $\phi : C_* \to D_*$ be a morphism of chain complexes both concentrated in positive degrees (i.e. with $H_i(C_*) = 0$ for $i \leq 0$. If the morphism ϕ is an $\mathbb{A}^1$-quasi-isomorphism, that is to say , then for any $s \geq 1$ the morphism of spaces*

$$L_s(\phi) : L_s(KC_*) \to L_s(KD_*)$$

is an $\mathbb{A}^1$-equivalence (of spaces).

For this reason the chain complexes (or spaces) of the form $L_{\mathbb{A}^1}$ $(L_s(KC_*))$ are called the total derived functors of the (non-additive) functor L_s.

Thus it follows that the above space $F_s^{\mathbb{A}^1} \cong L_{\mathbb{A}^1}(L_{s-1}(K(\tilde{C}_*(\mathcal{X})[-1])))$ is simplicially weakly equivalent to

$$F_s^{\mathbb{A}^1} \cong L_{\mathbb{A}^1}(L_{s-1}(K(\tilde{C}_*^{\mathbb{A}^1}(\mathcal{X})[-1])))$$

The right hand side is called the total $\mathbb{A}^1$-derived functor of L_{s-1} on $\tilde{C}_*^{\mathbb{A}^1}$ $(\mathcal{X})[-1])$, see below for some properties of these.

Together with what we have done before, we may summarize the previous results as follows. Let $\mathcal{X}$ be a 1-connected $\mathbb{A}^1$-local reduced pointed space. For each $s \geq 1$ set

$$W_s(\mathcal{X}) := B(L_{\mathbb{A}^1}(G(\mathcal{X})/\Gamma_s(G(\mathcal{X})))) \cong L_{\mathbb{A}^1}(B(G(\mathcal{X})/\Gamma_s(G(\mathcal{X}))))$$

and $\gamma_s(\mathcal{X}) := L_{\mathbb{A}^1}B\Gamma_s(G(\mathcal{X})))$ the homotopy fiber of the canonical morphism $\mathcal{X} \to W_s(\mathcal{X})$ (in the simplicial homotopy category).

THEOREM 3.7. *Let $\mathcal{X}$ be a 1-connected $\mathbb{A}^1$-local reduced pointed space. Then for any $s \geq 1$ the $\mathbb{A}^1$-local space $\gamma_s(\mathcal{X})$ is $\{1 + log_2(s)\}$-connected and in particular the morphism*

$$\mathcal{X} \to W_s(\mathcal{X})$$

is $\{1 + log_2(s)\}$-$\mathbb{A}^1$-connected.

Moreover the morphism $W_s \to W_{s-1}$ a principal fibration with group

$$BF_s^{\mathbb{A}^1} \cong BL_{\mathbb{A}^1}(L_{s-1}(K(\tilde{C}_*(\mathcal{X})[-1]))) \cong BL_{\mathbb{A}^1}(L_{s-1}(K(\tilde{C}_*^{\mathbb{A}^1}(\mathcal{X})[-1])))$$

Clearly it follows:

COROLLARY 3.8. *Let $\mathcal{X}$ be a 1-connected $\mathbb{A}^1$-local pointed space. If for each $s \geq 1$ the space $BL_{\mathbb{A}^1}(L_{s-1}(K(\tilde{C}_*^{\mathbb{A}^1}(\mathcal{X})[-1])))$ has the rigidity property, then $\mathcal{X}$ has the rigidity property.*

The rigidity property of the space $BL_{\mathbb{A}^1}(L_{s-1}(K(\tilde{C}_*^{\mathbb{A}^1}(\mathcal{X})[-1])))$ is a property concerning only the chain complex $\tilde{C}_*^{\mathbb{A}^1}(\mathcal{X})$. We thus have reduced the rigidity property for such 1-connected $\mathbb{A}^1$-local pointed spaces $\mathcal{X}$ to a property involving only the reduced $\mathbb{A}^1$-chain complex $\tilde{C}_*^{\mathbb{A}^1}(\mathcal{X})$. In the next section we make this observation quite precise.

3.2. $\mathbb{A}^1$-homological characterization of the rigidity property.

$\mathbb{A}^1$-Derived functor of the free Lie algebra functor. Our aim here is to prove the following Theorem, mentioned has Theorem 10 in the introduction:

THEOREM 3.9. *Let $\mathcal{X}$ be a pointed simply-connected $\mathbb{A}^1$-local space over the base field k, ℓ be a prime different from $char(k)$. If the $\mathbb{A}^1$-chain complexes $C_*^{\mathbb{A}^1}(\mathcal{X}^n;\mathbb{Z}/\ell)$, $n \geq 1$, have the rigidity property at ℓ, then the space $\mathcal{X}$ satisfies the rigidity property at ℓ.*

To explain the fineness of the previous Theorem 3.9 let us make some observations.

REMARK 3.10. We already observed that there exists an $\mathbb{A}^1$-localization functor $L_{\mathbb{A}^1}$ on spaces which commutes to finite products, thus takes simplicial sheaves of $\mathbb{Z}/\ell$-modules to simplicial sheaves of $\mathbb{Z}/\ell$-modules. This follows from [**44**] Theorem 1.66 p. 69 and Lemma 3.20 p. 93.

Given a chain complex of sheaves of abelian groups C_* concentrated in non negative degrees, the associated morphism of normalized chain complexes $C_* \to (L_{\mathbb{A}^1}(KC_*))_*^N$ is an $\mathbb{A}^1$-quasi-isomorphism, but in general $(L_{\mathbb{A}^1}(KC_*))_*^N$ is not an $\mathbb{A}^1$-local complex in the sense of [**41**].

It is a non-trivial result [**41**, Corollary 3.24] that for a complex $C_* \in \mathcal{C}_*^{\geq 0}(k)$ which is 0-connected, the morphism

$$KC_* \to K(L_{\mathbb{A}^1}^{ab} C_*)$$

is an $\mathbb{A}^1$-equivalence of spaces, which implies that the chain $(L_{\mathbb{A}^1}(KC_*))_*^N$ is in that case canonically isomorphic to $L_{\mathbb{A}^1}^{ab}(C_*)$ in $D(\mathcal{A}b(k))$. Quickly said, we have a simplicial weak-equivalence

$$L_{\mathbb{A}^1}(KC_*) \cong K(L_{\mathbb{A}^1}^{ab} C_*)$$

This is false in general if C_* is not assumed to be 0-connected. For instance take $C_* = \mathbb{Z}(\mathbb{G}_m)$. Then $\mathbb{Z}(\mathbb{G}_m) = K(\mathbb{Z}(\mathbb{G}_m)) \to K(L_{\mathbb{A}^1}^{ab}\mathbb{Z}(\mathbb{G}_m))$ is not an $\mathbb{A}^1$-equivalence. This follows from the computations of *loc. cit.* and the fact that the sheaf (and thus the space) $\mathbb{Z}(\mathbb{G}_m)$ is $\mathbb{A}^1$-invariant and thus $\mathbb{A}^1$-local. If the previous morphism were an $\mathbb{A}^1$-equivalence it would thus be a simplicial weak-equivalence (between two $\mathbb{A}^1$-local spaces) but from [**41**, Theorem 2.37], the sheaf $\pi_0^{\mathbb{A}^1}$ of the space $K(L_{\mathbb{A}^1}^{ab}\mathbb{Z}(\mathbb{G}_m))$ is the sheaf $\underline{\mathbf{K}}_1^{MW}$ and the induced morphism in degree 0, $\mathbb{Z}(\mathbb{G}_m) \to \underline{\mathbf{K}}_1^{MW}$ is not an isomorphism. □

REMARK 3.11. Let $\mathcal{X}$ be a space. The canonical morphism of chain complexes

$$C_*(\mathcal{X}) \to C_*^{\mathbb{A}^1}(\mathcal{X})$$

from the simplicial chain complex to the $\mathbb{A}^1$-local chain complex of $\mathcal{X}$ is not in general a quasi-isomorphism. Of course by the very definition it is an $\mathbb{A}^1$-quasi isomorphism, as $C_*^{\mathbb{A}^1}(\mathcal{X})$ is the $\mathbb{A}^1$-localization of $C_*(\mathcal{X})$. But even if $\mathcal{X}$ is $\mathbb{A}^1$-local, $C_*(\mathcal{X}) \to C_*^{\mathbb{A}^1}(\mathcal{X})$ is also alsmost never a quasi-isomorphism! In other words in general, the homology sheaves of $C_*(\mathcal{X})$ will not be strictly $\mathbb{A}^1$-invariant even if $\mathcal{X}$ itself is $\mathbb{A}^1$-local.

For instance take $\mathcal{X}$ to be $B\mathbb{G}_m$, an $\mathbb{A}^1$-local space. Then $H_1(B\mathbb{G}_m) = \mathbb{G}_m$ and the morphism $C_*(B\mathbb{G}_m) \to C_*^{\mathbb{A}^1}(B\mathbb{G}_m)$ is an isomorphism in homology up to degree 1 included. But $H_2(B\mathbb{G}_m) = \Lambda_2(\mathbb{G}_m)$ can be shown not to be a strictly $\mathbb{A}^1$-invariant sheaf; as a consequence, the previous morphism of chain complexes is not an isomorphism on H_2.

One should also notice that a product of finitely many spaces whith the rigidity property has the rigidity property, and a (derived) tensor product of finitely many chain complexes with the rigidity property has the rigidity property, as it follows from the Künneth formula. However, a tensor product of finitely many $\mathbb{A}^1$-local chain complexes will not be in general $\mathbb{A}^1$-local.

We end up this remark by the following questions: let $\mathcal{X}$ be a space with the rigidity property is it true that its $\mathbb{A}^1$-localization $L_{\mathbb{A}^1}(\mathcal{X})$ also has the rigidity property ? Let C_* be a chain complex with the rigidity property. Is it true that its $\mathbb{A}^1$-localization $L^{ab}_{\mathbb{A}^1}(C_*)$ will have the rigidity property ? □

Now we sketch the proof of Theorem 3.9. We first observe that to prove Theorem 3.9 it suffices to establish:

THEOREM 3.12. *Let C_* be a chain complex of $\mathbb{Z}/\ell$-modules such that $H_i(C_*) = 0$ for $i \leq 0$ and such that for each integer $n \geq 1$ the $\mathbb{A}^1$-localization $L^{ab}_{\mathbb{A}^1}((C_*)^{\otimes n})$ has the rigidity property at ℓ. Then for each $s \geq 1$ the space*

$$L_{\mathbb{A}^1}(L_s(K(C_*))$$

has the rigidity property at ℓ.

Proof of Theorem 3.9 using Theorem 3.12. By Corollary 3.8 it is sufficient to show that the spaces $BL_{\mathbb{A}^1}(L_s(K(\tilde{C}^{\mathbb{A}^1}_*(\mathcal{X})[-1])))$ have the rigidity property at ℓ, or equivalently that the spaces $L_{\mathbb{A}^1}(L_s(K(\tilde{C}^{\mathbb{A}^1}_*(\mathcal{X})[-1])))$ have the rigidity property at ℓ by Remark 2.5. By Theorem 3.6, we know that

$$L_{\mathbb{A}^1}(L_s(K(\tilde{C}_*(\mathcal{X})[-1]))) \to L_{\mathbb{A}^1}(L_s(K(\tilde{C}^{\mathbb{A}^1}_*(\mathcal{X})[-1])))$$

is a simplicial weak equivalence, as well as

$$L_{\mathbb{A}^1}(L_s(K(\tilde{C}_*(\mathcal{X};\mathbb{Z}/\ell)[-1]))) \to L_{\mathbb{A}^1}(L_s(K(\tilde{C}^{\mathbb{A}^1}_*(\mathcal{X};\mathbb{Z}/\ell)[-1])))$$

Now we see that it suffices to check that the spaces $L_{\mathbb{A}^1}(L_s(K(\tilde{C}^{\mathbb{A}^1}_*(\mathcal{X};\mathbb{Z}/\ell)[-1])))$ have the rigidity property at ℓ. Now this follows from Theorem 3.12 applied to the chain complex $C_* = \tilde{C}^{\mathbb{A}^1}_*(\mathcal{X};\mathbb{Z}/\ell)[-1])$; observe that the $\mathbb{A}^1$-localization of the chain complexes $(\tilde{C}^{\mathbb{A}^1}_*(\mathcal{X};\mathbb{Z}/\ell)[-1]))^{\otimes n}$ is quasi-isomorphic to the chain complex $\tilde{C}^{\mathbb{A}^1}_*(\mathcal{X}^{\wedge n};\mathbb{Z}/\ell)[-n])$, which, being a summand of $C^{\mathbb{A}^1}_*(\mathcal{X}^n;\mathbb{Z}/\ell)[-n])$ has the rigidity property by assumptions, which shows that C_* satisfies the assumption of Theorem 3.12. □

Now to get Theorem 3.12 we use some standard technics of devissage together with the following remarks.

From [**9, 51**] the free Lie algebra of a direct sum $V_1 \oplus V_2$ decomposes canonically as:

$$\oplus_{I\in\mathcal{B}} L_*(V_I) \cong L_*(V_1 \oplus V_2)$$

where the index set $\mathcal{B}$ is the set of basic commutators $\mathcal{B}$ in two letters. This means in particular that the above isomorphism is an isomorphism of functors. This fact allows one for instance to control the statement of Theorem 3.12 if we have a cofibration sequence of chain complexes.

Let us recall some notations for functors $\mathcal{A}b \to \mathcal{A}b$. T_n will denote the n-th tensor product functor $A \mapsto A^{\otimes n}$, S_n will denote the n-th symmetric tensor product functor $A \mapsto A^{\otimes n}/_{\Sigma_n}$. We will also have to use the ℓ-cyclic product functor $A \mapsto C_\ell(A) := A^{\otimes \ell}/_{\mathbb{Z}/\ell}$ and more generally for $H \subset S_m$ a subgroup, we denote by S_m^H the functor $A \mapsto A^{\otimes n}/_H$. Observe that if $[S_m : H]$ is prime to ℓ, the epimorphism $S_m^H \to S_m$ admits a section when restricted (as functor) to the category of $\mathbb{Z}/\ell$-modules. We extend these notations in the obvious way to sheaves of abelian groups.

The remaining idea to prove the Theorem 3.12 is the following. Let C_* be a chain complex of $\mathbb{Z}/\ell$-modules such that $H_i(C_*) = 0$ for $i \leq 0$ and such that for each integer $n \geq 1$ the $\mathbb{A}^1$-localization $L_{\mathbb{A}^1}^{ab}((C_*)^{\otimes n})$ has the rigidity property. We deduce from this that the $\mathbb{A}^1$-localization $L_{\mathbb{A}^1}^{ab}(S_n((C_*))$ have the rigidity property by some standard homological algebra step. One proceed by induction on n. Then one observe that if ℓ divides n the functor S_n can be expressed in term of functors $S_{n'}$ with $n' < n$ and the cyclic functor C_ℓ. This one is understood in terms of the cohomology of $\mathbb{Z}/\ell$ with values in T_ℓ.

Now we still have to deduce from these fact the rigidity property for the $L_{\mathbb{A}^1}^{ab}(L_n(C_*))$'s. The last idea is to use the Poincaré-Birkhoff-Witt Theorem. For M a $\mathbb{Z}/\ell$-module, the isomorphism $L_1(M) = T_1(M) = M$ extends to a canonical morphism $L_*(M) \to T_*(M)$, obtained by considering the tensor algebra as a Lie algebra. This morphism is injective and more precisely, let $U_i(M)_* \subset T_*(M)$ be the sub-$\mathbb{Z}/\ell$-module generated by the products of j-elements of $L_*(M)$ for each $j \leq i$. One thus gets an increasing filtration of $\mathbb{N}$-graded $\mathbb{Z}/\ell$-modules

$$L_*(M) = U_1(M)_* \subset U_2(M)_* \subset \cdots \subset U_i(M)_* \subset \cdots \subset T_*(M) \tag{3.3}$$

The Poincaré-Birkhoff-Witt Theorem [**4**] asserts that the canonical morphism of commutative graded (by i) algebras $\oplus_i S_i(L_*(M)) \to \oplus_i U_i(M)_*/U_{i-1}(M)_*$ is an isomorphism.

In a given degree s this gives an increasing filtration

$$L_s(M) = U_1(M)_s \subset U_2(M)_s \subset \cdots \subset U_{s-1}(M)_s \subset U_s(M) = T_s(M) \tag{3.4}$$

such that the quotient $U_i(M)_s/U_{i-1}(M)$ is canonically isomorphic to the degree i part of the symmetric algebra on the free Lie algebra over M. Using the usual formula $S_*(M \oplus N) = S_*(M) \otimes S_*(N)$, we see that for $i \geq 2$, $U_i(M)_s/U_{i-1}(M)_s$ is as functor equal to the finite direct sum of the functors of the form $M \mapsto S_{j_1}(L_{t_1}(M)) \otimes \cdots \otimes S_{j_a}(L_{t_a}(M))$ with $\Sigma j_b t_b = s$ and $\Sigma j_b = i$.

This permits to finish the proof of Theorem 3.12 by induction on s. More details may be found in [**42**]. □

4. Reduction to the rigidity property for the $C_*^{\mathbb{A}^1}((\mathbb{G}_m)^{\wedge n})$'s

4.1. $\mathbb{A}^1$-chain complex of mixed Tate type.

DEFINITION 4.1. A full subcategory $\mathcal{C} \subset D_{\mathbb{A}^1}^{\geq 0}(k)$ is called a *Class of complexes* if the following properties hold:

(1) $\mathcal{C}$ is stable under arbitrary direct sums (and thus contains the 0 complex);

(2) For any exact triangle $A \to B \to C \to A[1]$ in $D^{\geq 0}_{\mathbb{A}^1}(k)$, if $A \in \mathcal{C}$ and $(B \in \mathcal{C})$ then $(C \in \mathcal{C})$.

The class $\mathcal{C}$ is said to be multiplicative if moreover:

(3) For any $A \in \mathcal{C}$ and $(B \in \mathcal{C})$ then $A \otimes B \in \mathcal{C}$.

We will say that a full subcategory $\mathcal{C} \subset D^{\geq 0}_{\mathbb{A}^1}(k)$ is a *strict Class of complexes* if the following axiom holds in place of (1):

(1') $\mathcal{C}$ is stable under finite direct sums (and thus contains the 0 complex) and by direct summand.

Observe that a class is stable by telescopes and in particular by direct summand. Also classes and strict classes are stable by suspension.

EXAMPLE 4.2. If $\mathcal{C} \subset \mathcal{A}b_{\mathbb{A}^1}(k)$ is a Serre class in the abelian category $\mathcal{A}b_{\mathbb{A}^1}(k)$ in the usual sense then the full subcategory $D^{\mathcal{C}} \subset D^{\geq 0}_{\mathbb{A}^1}(k)$ consisting of chain complex with homology objects in $\mathcal{C}$ is a class of complexes in the previous sense.

EXAMPLE 4.3. The full subcategory $D_{\mathbb{A}^1}(\mathcal{A}b(k)) \subset D(\mathcal{A}b(k))$ consisting of $\mathbb{A}^1$-local chain complexes is a class of complex. Observe however it is not multiplicative.

EXAMPLE 4.4. An intersection of a family of classes is also a class of complexes.

Let $(C_*(i))_i$ be a family of chain complexes in $D^{\geq 0}_{\mathbb{A}^1}(k)$. We let $\mathcal{C}(\{C_*(i)\}_i) \subset D^{\geq 0}_{\mathbb{A}^1}(k)$ be the class generated by the $C_*(i)$, i.e. the intersection of all classes of complexes which contains this family. We may also introduce the strict class of complexes generated by a family of complexes.

Define by induction on n an increasing family of full subcategories $\mathcal{C}^{(0)} \subset \mathcal{C}^{(1)} \subset \dots \mathcal{C}^{(n)} \dots \subset D^{\geq 0}_{\mathbb{A}^1}(k)$ as follows: $\mathcal{C}^{(0)}$ is the full subcategory with objects the elements of the family $(C_*(i))_i$. The category $\mathcal{C}^{(n)}$ has for objects the factors of the objects of $\mathcal{C}^{(n-1)}$ and the cones of morphisms in $\mathcal{C}^{(n-1)}$. Clearly

$$\bigcup_n \mathcal{C}^{(n)} = \mathcal{C}(\{C_*(i)\}_i) \subset D^{\geq 0}_{\mathbb{A}^1}(k)$$

EXAMPLE 4.5. If $\mathcal{C} \subset D^{\geq 0}(\mathcal{A}b(k))$ is a class of complexes, the full subcategory $\mathcal{C}_{\mathbb{A}^1} \subset D^{\geq 0}(\mathcal{A}b(k))$ consisting of complexes C_* whose $\mathbb{A}^1$-localization $L^{ab}_{\mathbb{A}^1}(C_*)$ is in $\mathcal{C}$ is a class of complexes: this follows easily from the fact that $L^{ab}_{\mathbb{A}^1}(-)$ is exact (and preserves connectivity).

EXAMPLE 4.6. The full subcategory $\mathcal{C}^{rig,\ell,\geq 0} \subset D^{\geq 0}(\mathcal{A}b(k))$ consisting of $\mathbb{A}^1$-local chain complexes which have the rigidity property at ℓ is a class of complexes. This follows from Lemma 2.4.

It follows from Example 4.5 that the full subcategory $\mathcal{C}^{rig,\ell,\geq 0}_{\mathbb{A}^1} \subset D^{\geq 0}(\mathcal{A}b(k))$ consisting of chain complexes whose $\mathbb{A}^1$-localization has the rigidity property at ℓ is a class of complexes.

The usual splitting of the sheaf $\mathbb{Z}((\mathbb{G}_m)^n)$ as

$$\mathbb{Z}((\mathbb{G}_m)^n) \cong \oplus_{I \subset \{1,\ldots,n\}} \mathbb{Z}(\mathbb{G}_m^{\wedge |I|}) \tag{4.1}$$

induces a corresponding splitting of the $\mathbb{A}^1$-chain complex $C_*^{\mathbb{A}^1}((\mathbb{G}_m)^n)$ as a direct sum of $C_*^{\mathbb{A}^1}((\mathbb{G}_m)^{\wedge n})$'s. Consequently, the class of complexes generated by the $C_*^{\mathbb{A}^1}((\mathbb{G}_m)^{\wedge i})\}_i)$'s is the same as the one generated by the $C_*^{\mathbb{A}^1}((\mathbb{G}_m)^i)\}_i)$'s.

DEFINITION 4.7. The class of *mixed Tate complexes* is the class

$$\mathcal{C}_{\mathbb{A}^1}^{Tate}(k) := \mathcal{C}(\{C_*^{\mathbb{A}^1}((\mathbb{G}_m)^{\wedge i})\}_i) = \mathcal{C}(\{C_*^{\mathbb{A}^1}((\mathbb{G}_m)^i)\}_i)$$

generated by the $\mathbb{A}^1$-chain complexes of all the smash-power of $\mathbb{G}_m$'s or equivalently by all the split tori.

The class of *strict mixed Tate complexes* is the strict class $\mathcal{C}_{\mathbb{A}^1}^{str.-Tate}(k)$ generated by the $\mathbb{A}^1$-chain complexes of all the smash-power of $\mathbb{G}_m$'s.

Observe that $\mathcal{C}_{\mathbb{A}^1}^{str.-Tate}(k) \subset \mathcal{C}_{\mathbb{A}^1}^{Tate}(k)$ are both multiplicative and contained in the class of $\mathbb{A}^1$-local chain complexes.

4.2. Bruhat decomposition and $\mathbb{A}^1$-chain complex of G. In this section we deal with:

THEOREM 4.8. *Let G be a split semi-simple k-group. Then the $\mathbb{A}^1$-chain complexes $C_*^{\mathbb{A}^1}(G)$ of G is of strict mixed Tate type and the $\mathbb{A}^1$-chain complex $C_*^{\mathbb{A}^1}(BG)$ of BG is the suspension of a mixed Tate complex.*

If furthermore G is simply connected, then $C_*^{\mathbb{A}^1}(G)$ is the suspension of a strict mixed Tate complex.

PROOF. The first statement follows from the Bruhat decomposition of G. Let $T \subset G$ be a split maximal torus, $T \subset B \subset G$ a Borel subgroup containing T. There is an increasing filtration of open subsets [**55**]

$$\emptyset = \Omega_{s+1} \subset \Omega_s \subset \Omega_{s-1} \subset \cdots \subset \Omega_0 = G$$

and for each i an isomorphism of schemes

$$\Omega_{i-1} - \Omega_i \cong \amalg_{w \in W | length(w) = i} \mathbb{A}^i \times B$$

where $length()$ is the length function on the Weyl group W of G, and s is the maximal possible length. Observe that there is only one element of length s and that the "big cell" Ω_s is an open subscheme isomorphic to $\mathbb{A}^s \times B$.

By the theorem of Quillen-Suslin, and its generalization to rings of Laurent polynomials [**32**, Corollary 4.9 p. 146], the normal bundle of the closed immersion $\Omega_{i-1} - \Omega_i \to \Omega_{i-1}$ is trivial and of rank $s - i$ (remember that as a k-variety B is the product of T and an affine space [**55**].

Now using the homotopy purity theorem of [**44**], the quotients sheaves Ω_{i-1}/Ω_i are canonically isomorphic in the $\mathbb{A}^1$-homotopy category to the Thom space of the trivial bundle, which means a pointed space of the form

$S^{s-i} \wedge (\mathbb{G}_m)^{\wedge s-i} \wedge (\mathbb{A}^i \times B_+)$. As $T \to B$ is an $\mathbb{A}^1$-homotopy equivalence, it follows that the quotient sheaves Ω_{i-1}/Ω_i are indeed isomorphic in the pointed $\mathbb{A}^1$-homotopy category to $S^{s-i} \wedge (\mathbb{G}_m)^{\wedge s-i} \wedge (T_+)$.

It follows at once that the $\mathbb{A}^1$-chain complex G is of mixed Tate type. The statement concerning $C_*^{\mathbb{A}^1}(BG)$ follows from this, the fact that the class of mixed Tate complexes is multiplicative, by considering the usual skeletal filtration of BG, $* = F^0(BG) \subset F^1(BG) = S^1 \wedge G \subset \cdots \subset F^s(BG) \subset \ldots BG$, whose quotients are $F^s(BG)/F^{s-1}(BG) \cong S^i \wedge G^{\wedge i}$.

To prove the last statement, we observe by the Bruhat decomposition that the cone C of $T \subset G$ (do no confound with the homogeneous variety G/T) is always the suspension of a strict mixed Tate type.

Now we saw in Theorem 1.5 that if G is simply connected it is $\mathbb{A}^1$-connected. It follows from the results in [**41**] that the morphism $T \to G$ is then trivial in the pointed $\mathbb{A}^1$-homotopy category. Thus the cone C is isomorphic to the wedge $G \vee \Sigma(T)$ which proves the last property. □

REMARK 4.9. Observe that because $BG \to BSing_\bullet^{\mathbb{A}^1}(G)$ is an $\mathbb{A}^1$-weak equivalence, it follows that the morphism of chain complexes

$$C_*^{\mathbb{A}^1}(BG) \to C_*^{\mathbb{A}^1}(BSing_\bullet^{\mathbb{A}^1}(G))$$

is a quasi-isomorphism. □

It follows from all what we have been doing so far, that for G a split semi-simple and simply connected algebraic k-group, to check that the pointed $\mathbb{A}^1$-local simply connected space

$$BSing_\bullet^{\mathbb{A}^1}(G)$$

has the rigidity property at ℓ it suffices to prove that for each $n \geq 1$, the chain complex $C_*^{\mathbb{A}^1}((\mathbb{G}_m)^{\wedge n}; \mathbb{Z}/\ell)$ has the rigidity property. This follows from Remark 4.6, Theorem 3.9 and Theorem 4.8.

Thus Theorem 3 for G is a consequence of the following Theorem, which by the way, doesn't depend on G. It is proven in Section 5 below.

THEOREM 4.10. *Assume k is a prime field*[6]. *For any $n \geq 0$ the chain complex*

$$L_{\mathbb{A}^1}^{ab}\mathbb{Z}((\mathbb{G}_m)^{\wedge n}) = C_*^{\mathbb{A}^1}(\mathbb{G}_m^{\wedge n})$$

has the rigidity property.

REMARK 4.11. We already mentioned that in fact we can prove the strong rigidity property for these complexes. See [**43**].

[6]of characteristic $\neq 2$

Some further questions concerning the $C_*^{\mathbb{A}^1}((\mathbb{G}_m)^{\wedge n})$'s. We now mention some concrete consequences of the Theorem 4.10.

COROLLARY 4.12. *For each $n \geq 1$ the chain complex $C_*^{\mathbb{A}^1}((\mathbb{G}_m)^{\wedge n};\mathbb{Z}/\ell)$ is locally constant in the étale topology and thus the canonical morphism between $\mathbb{A}^1$-local complexes*

$$C_*^{\mathbb{A}^1}((\mathbb{G}_m)^{\wedge n};\mathbb{Z}/\ell) \to \mathbb{Z}/\ell(n)[n]$$

is an isomorphism in the étale topology, where the complex $\mathbb{Z}/\ell(n)$ is the mod ℓ motivic complex in weight n defined by Suslin-Voevodsky [**67**].

PROOF. Theorem 4.10 implies that $C_*^{\mathbb{A}^1}((\mathbb{G}_m)^{\wedge n};\mathbb{Z})$ and thus $C_*^{\mathbb{A}^1}((\mathbb{G}_m)^{\wedge n};\mathbb{Z}/\ell)$ have the rigidity property at ℓ. This is also true for $\mathbb{Z}/\ell(n)[n]$ by [**67**, Proposition 3.3.3 p. 215]. Thus both complexes are locally constant in the étale topology. Thus it suffices to prove that working over an algebraic closure F of k, and given M a constant sheaf of $\mathbb{Z}/\ell$-vector spaces over Sm_F, the morphisms in the derived category of étale sheaves of $\mathbb{Z}/\ell$-vector spaces on Sm_k to the shifts $M[*]$ from both complexes are equal. The group of morphisms from $C_*^{\mathbb{A}^1}((\mathbb{G}_m)^{\wedge n};\mathbb{Z}/\ell)$ to $M[*]$ in the étale topology over F is by definition $\tilde{H}^*_{et}((\mathbb{G}_m)_F^{\wedge n};M)$. By [**67**, Proposition 3.3.3 p. 215] the complex $\mathbb{Z}/\ell(n)[n]$ is over F isomorphic to $(\mu_\ell)^{\otimes n}[n]$ in the étale topology. We see that we get the same answer by the usual computation of étale cohomology of $\mathbb{G}_m^n$. □

In fact we conjecture the following. Let $\underline{\mathbf{K}}_n^{MW}$ be the sheaf of unramified Milnor-Witt K-theory defined in [**41**]. There is a canonical epimorphism $\underline{\mathbf{K}}_n^{MW} \to \underline{\mathbf{K}}_n^{M}$ and we let $\mathbb{Z}^{\mathbb{A}^1}(n)$ be the chain complex of sheaves of abelian groups on Sm_k in the Nisnevich topology defined as the fiber product

$$\mathbb{Z}^{\mathbb{A}^1}(n) := \mathbb{Z}(n) \times_{\underline{\mathbf{K}}_n^{M}[-n]} \underline{\mathbf{K}}_n^{MW}[-n]$$

From [**41**] there exists a canonical morphism

$$C_*^{\mathbb{A}^1}(\mathbb{G}_m^{\wedge n}) \to \mathbb{Z}^{\mathbb{A}^1}(n)[n]$$

which induces the isomorphism $H_0^{\mathbb{A}^1}(\mathbb{G}_m^{\wedge n}) = \underline{\mathbf{K}}_n^{MW}$ on the 0 homology sheaves.

The following conjecture is a refinement of Corollary 4.12:

CONJECTURE 4.13. *For any field k, the morphism*

$$C_*^{\mathbb{A}^1}(\mathbb{G}_m^{\wedge n}) \to \mathbb{Z}^{\mathbb{A}^1}(n)[n]$$

is a quasi-isomorphism.

Indeed the fiber of the above morphism is a complex concentrated in degree 0, equal by [**41**] and the Milnor conjecture to the sheaf $\underline{\mathbf{I}}^{n+1}$ by [**37**]; now the sheaves $\underline{\mathbf{I}}^n$ are zero in the étale topology for $n \geq 1$.

5. The rigidity property for the $C_*^{\mathbb{A}^1}((\mathbb{G}_m)^{\wedge n})$'s

In this section we prove that the chain complexes $C_*^{\mathbb{A}^1}((\mathbb{G}_m)^{\wedge n})$ have the rigidity property, completing thus the proof of Theorem 3 and Corollary 6.

Our method is to endow the homology sheaves of $C_*^{\mathbb{A}^1}((\mathbb{G}_m)^{\wedge n})$ with a natural structure of sheaf with transfers in a slightly more general (or weaker) sense as the notion of Voevodsky [**67**]. In Section 5.2 below we achieve the proof of Theorem 3 by showing that this new kind of sheaves have the rigidity property at ℓ by reducing to the classical rigidity theorem from [**63**]. These structure on the homology sheaves are coming in fact from transfers morphisms at the chain complexes level:

$$RC_*^{\mathbb{A}^1}((\mathbb{G}_m)^n)^L \to RC_*^{\mathbb{A}^1}((\mathbb{G}_m)^n)^K$$

where $K \subset L$ is a monogenous finite extension between finite type fields extension of k, and the notation RC_*^S denotes the derived internal functional complex. This is done in details in [**43**].

Remark 5.1. Contrary to our first belief, the construction of the above transfer morphisms on the $\mathbb{A}^1$-localization of the $\mathbb{Z}((\mathbb{G}_m)^{\wedge n})$ are not directly related to the transfers morphisms on the sheaf $\mathbb{G}_m^n$. It is not so clear to formalize our construction and to write down a minimal list of axioms for a sheaf M that guaranties that the homology sheaves of the chain complex $L_{\mathbb{A}^1}^{ab}(\mathbb{Z}(\mathcal{M}))$ will have this type of structure. It seems that a good general candidate for is a sheaf of the form $\mathbb{Z}(\mathbb{G}_m) \otimes M$ where M is any reasonable sheaf. This is still an open question.

On the converse, even if $\mathcal{M}$ be a strictly $\mathbb{A}^1$-invariant sheaf with transfers in the sense of Voevodsky then in general the chain complex $L_{\mathbb{A}^1}^{ab}(\mathbb{Z}(\mathcal{M}))$ has not the rigidity property at ℓ. In particular it has no suitable transfers structure in general. For instance take the free sheaf of abelian groups $\mathbb{Z}(E)$ on the sheaf with transfers $\mathcal{M} = E$ given by an elliptic curve E over k. Then the sheaf $\mathbb{Z}(E)$ itself is a strictly $\mathbb{A}^1$-invariant sheaf, and is thus equal to its own $\mathbb{A}^1$-localization. One may check that it has not the rigidity property at ℓ. □

5.1. The Transfers morphisms on the $\mathbb{A}^1$-localization of the $\mathbb{Z}(\mathbb{G}_m^{\wedge n})$'s. We consider the free pointed sheaf of abelian groups $\mathcal{Z}(n) := \mathbb{Z}(\mathbb{G}_m^{\wedge n})$ on the smash-product $\mathbb{G}_m^{\wedge n}$ (this means that we impose the relation "base point" $= 0$). We define below an explicit inclusion of sheaves of abelian groups on Sm_k of the form:

$$0 \to \mathcal{Z}(n) \subset \mathcal{Z}_{tr}(n)$$

with two basic properties:

1) For each monogenous finite extension $K \subset L$ of fields of finite type over k, there is a canonical transfer morphism

$$\mathcal{Z}(n)(L) \to \mathcal{Z}_{tr}(n)(K)$$

with nice natural properties; $\mathcal{Z}_{tr}(n)$ is more or less the universal sheaf with this property, it is defined as the sheaf of finite correspondences to $\mathbb{A}^1$ with coefficients in $\mathcal{Z}(n)$ in the spririt of [**67**].

2) The inclusion $\mathcal{Z}(n) \to \mathcal{Z}_{tr}(n)$ is is an $\mathbb{A}^1$-equivalence, in a quite explicit way. In particular applying the $\mathbb{A}^1$-localization functor to $\mathcal{Z}(n) \to \mathcal{Z}_{tr}(n)$ yields a quasi isomorphism of the form:

$$C_*^{\mathbb{A}^1}(\mathbb{G}_m^{\wedge n}) = L_{\mathbb{A}^1}^{ab}(\mathcal{Z}(n)) \cong L_{\mathbb{A}^1}^{ab}(\hat{\mathcal{Z}}(n))$$

The transfers morphisms from 1) and some general fact on $\mathbb{A}^1$-localization of derived functional complexes allow one to construct the transfers morphisms on each of the homology sheaves $H_*^{\mathbb{A}^1}(\mathbb{G}_m^{\wedge n})$ of $C_*^{\mathbb{A}^1}(\mathbb{G}_m^{\wedge n})$ which will endow these homology sheaves with a structure of *strictly* $\mathbb{A}^1$*-invariant sheaves with generalized transfers* that we describe and use in the next Section 5.2 below, in which we also prove a rigidity theorem.

REMARK 5.2. The transfers morphisms we get are the "correct" ones. For instance the transfers we get in this way on $H_0^{\mathbb{A}^1}((\mathbb{G}_m)^{\wedge n})$ which is the unramified sheaf of Milnor-Witt K-theory in weight n, $\underline{\mathbf{K}}_n^{MW}$ described in [**41**], are the "usual" ones, and are indeed compatible with both the transfers in Milnor K-theory $\underline{\mathbf{K}}_n^M$ and with the canonical transfers in the n-th power of the fundamental ideal $\underline{\mathbf{I}}^n$ [**1**][**53**] through the canonical epimorphisms

$$\begin{array}{ccc} \underline{\mathbf{K}}_n^{MW} & \to & \underline{\mathbf{K}}_n^M \\ \downarrow & & \\ \underline{\mathbf{I}}^n & & \square \end{array}$$

In the sequel, we simply denote the sheaf $\mathcal{Z}(n)$ by $\mathcal{Z}$ when no confusion can arise on n, which is fixed everywhere from now on. The sheaf of abelian groups $\mathcal{Z}$ has many good properties. It is an unramified sheaf of abelian groups in the sense of [**41**], which means that its value on a smooth irreducible k-scheme X with function field K is the intersection in $\mathcal{Z}(K)$ of the $\mathcal{Z}(\mathcal{O}_{X,x})$ where x runs over the set $X^{(1)}$ of points of codimension 1 in X. It is moreover $\mathbb{A}^1$-invariant. However $\mathcal{Z}$ is not a strictly $\mathbb{A}^1$-invariant sheaf though the sheaf $\mathbb{G}_m^n$ has this property; this is the whole point of the story of course!

We start by observing that the sheaf $\mathcal{Z}$ admits a canonical structure of $\mathbb{Z}[\mathbb{G}_m]$-module. Here $\mathbb{Z}[\mathbb{G}_m]$ is the free sheaf of abelian groups on the sheaf of groups $\mathbb{G}_m$ (no condition on the base point) and as such is indeed a sheaf of commutative rings. We take the multiplication

$$\mathbb{Z}[\mathbb{G}_m] \times \mathbb{Z}((\mathbb{G}_m)^{\wedge n}) \to \mathbb{Z}((\mathbb{G}_m)^{\wedge n})$$

$$(U, \{\alpha_1, \dots, \alpha_n\}) \to \{U\alpha_1, \alpha_2, \dots, \alpha_n\} - \{U, \alpha_2 \dots, \alpha_n\}$$

(the correcting term is there to garanty that the map factors through the smash product.)

For any integer $n \in \mathbb{N}$, we set $[n]_\epsilon = \sum_{i=0}^{n-1}\{(-1)^i\}$ which we consider as an element of $\mathbb{Z}[\mathbb{G}_m](k) = \mathbb{Z}[k^\times]$. Observe that one has the easy formula

$$(5.1) \qquad [n]_\epsilon.[m]_\epsilon = [nm]_\epsilon$$

The above $\mathbb{Z}[\mathbb{G}_m]$-structure will *a priori* play a little role in the constructions below, only through the action of the $[n]_\epsilon$. However, to check the property of the transfers morphims it will be essential. See Section 5.2 below.

If X is a smooth k-scheme we let $\mathcal{C}(X;\mathbb{A}^1)$ be the set of points $y \in \mathbb{A}^1 \times X$ such that the closure $\overline{y}$ (an irreducible closed integral subscheme in $\mathbb{A}^1 \times X$) is finite over X and dominate an irreducible component of X (see [**67**]). We now define the sheaf $\mathcal{Z}_{tr}$. Its values on a smooth k-scheme X is the abelian group:

$$\mathcal{Z}_{tr}(X) := \oplus_{y\in\mathcal{C}(X,\mathbb{A}^1)}\mathcal{Z}(\overline{y})$$

One of the hidden points of this definition is that we may actually define $\mathcal{Z}(\overline{y})$, which means that $\mathcal{Z}$ as a presheaf can be extended to singular k-schemes of the type $\overline{y}$.

Observe also that if X is irreducible with function field K one has a canonical inclusion $\mathcal{Z}_{tr}(X) \subset \mathcal{Z}_{tr}(K)$. This follows from the fact that $Y \times_X Spec(K) \to Y$ induces an injection $\mathcal{Z}(Y) \subset \mathcal{Z}(Y \times_X Spec(K))$.

Now we turn this definition into a presheaf on Sm_k as follows. Let $f : X' \to X$ be a morphism in Sm_k. We want to define $\mathcal{Z}_{tr}(f) : \mathcal{Z}_{tr}(X) \to \mathcal{Z}(X')$. Clearly if $X = X_1 \amalg X_2$ the group $\mathcal{Z}_{tr}(X)$ splits accordingly as $\mathcal{Z}_{tr}(X_1) \oplus \mathcal{Z}_{tr}(X_2)$. To define $\mathcal{Z}_{tr}(f)$ we may thus assume that X' and X are irreducible.

Assume first that f is dominant. Let K be the function field of X and L that of X'. As $\mathcal{Z}_{tr}(X) \subset \mathcal{Z}_{tr}(K)$ and $\mathcal{Z}_{tr}(X') \subset \mathcal{Z}_{tr}(L)$ we need first to define the morphism $\mathcal{Z}_{tr}(K) \to \mathcal{Z}_{tr}(L)$ induced by the field extension $\phi : K \subset L$.

Given $y \in \mathcal{C}(Spec(K);\mathbb{A}^1) = (\mathbb{A}^1_K)^{(1)}$, we denote by $\{z|y\}$ the finite set of closed points in $\mathbb{A}^1_L$ which map to y through the induced morphism $\mathbb{A}^1_L \to \mathbb{A}^1_K$ and we write $z|y$ to mean $z \in \{z|y\}$. In other words, the set of such z is exactly the set of points in the closed subscheme $Spec(\kappa(y) \otimes_E F)$ of $\mathbb{A}^1_F$. For each $z|y$ we let $\lambda(z|y) \in \mathbb{N}$ denote the lenght of z in $\kappa(y) \otimes_E F$. Then we take for morphism $\mathcal{Z}_{tr}(K) \to \mathcal{Z}_{tr}(L)$ the direct sum indexed by the y's of the following homomorphism whose source is the factor $\mathcal{Z}(\kappa(y))$ of $\mathcal{Z}_{tr}(K)$ and the target the summand $\oplus_{z|y}\mathcal{Z}(\kappa(z))$ of $\mathcal{Z}_{tr}(L)$:

$$\sum_{z|y}[\lambda(z|y)]_\epsilon \,.\, \mathcal{Z}(\phi_z) : \mathcal{Z}(\kappa(y)) \to \oplus_{z|y}\mathcal{Z}(\kappa(z))$$

Here ϕ_z is the induced fields extension $\kappa(y) \subset \kappa(z)$ and $\mathcal{Z}(\phi_z) : \mathcal{Z}(\kappa(y)) \to \mathcal{Z}(\kappa(z))$ the induced group homomorphism.

One checks that this construction induces a functor $(\tilde{Sm}_k)^{op} \to \mathcal{A}b$ from smooth k-schemes and dominant morphisms to that of abelian groups. It remains to extends this to a functor $(Sm_k)^{op} \to \mathcal{A}b$ and for this we first define restriction morphisms

$$\mathcal{Z}_{tr}(X) \to \mathcal{Z}_{tr}(Z)$$

for any closed immersion $Z \subset X$ in Sm_k everywhere of codimension 1.

One may reduce to X being the localization of the generic point of Z, that is to say $X = Spec(A)$ where A is a geometric discrete valuation ring with residue field κ the function field of Z. In that case given $y \in \mathcal{C}(X, \mathbb{A}^1)$ we let μ_y be the length of the finite κ-scheme $\overline{y} \times_X Spec(\kappa)$. Let also $t(y) \in \mathcal{C}(Spec(\kappa), \mathbb{A}^1)$ the image of y in $\mathbb{A}^1_\kappa$. We take for morphism

$$\mathcal{Z}_{tr}(X) = \mathcal{Z}_{tr}(Spec(A)) \to \mathcal{Z}_{tr}(Spec(\kappa)) = \mathcal{Z}_{tr}(Z)$$

the direct sum over the y's of the morphisms of the form:

$$[\mu]_\epsilon : \mathcal{Z}(\overline{y}) \to \mathcal{Z}(\kappa(t(y)))$$

induced by the restriction morphisms on $\mathcal{Z}$. One may check:

THEOREM 5.3. *The previous construction extends uniquely to a presheaf of $\mathbb{Z}[\mathbb{G}_m]$-modules $\mathcal{Z}_{tr}$ on Sm_k. This presheaf is a sheaf in the Nisnevich topology. Moreover the natural transformation $\mathcal{Z} \to \mathcal{Z}_{tr}$ which corresponds to the zero section $0 \in \mathcal{C}(-, \mathbb{A}^1)$ is a monomorphism of sheaves (of $\mathbb{Z}[\mathbb{G}_m]$-modules).*

Now we prove that the inclusion $\mathcal{Z} \to \mathcal{Z}_{tr}$ is an $\mathbb{A}^1$-equivalence. The proof is completely explicit and proceeds in several steps (basically 3). First we observe that $\mathcal{Z}_{tr}$ contains a canonical subsheaf which we denote by $c(-; \mathbb{A}^1)\tilde{\otimes}\mathcal{Z}$ whose values on the henselization A of a point in a smooth k-scheme is equal to

$$c(Spec(A); \mathbb{A}^1) \otimes \mathcal{Z}(A) = \oplus_{y \in \mathcal{C}(Spec(A), \mathbb{A}^1)} \mathcal{Z}(A)$$

It is straithforward to check that these form a subsheaf of $\mathcal{Z}_{tr}$. The notation $c(-; \mathbb{A}^1)\tilde{\otimes}\mathcal{Z}$ means that the sheaf is not exactly the tensor product, as the restriction morphisms use the multiplication by the $[n]_\epsilon$'s at some point.

Now the obvious $\mathbb{A}^1$-homotopy on $c(-, \mathbb{A}^1)$ "multilplication by the scalar" which gives an $\mathbb{A}^1$-deformation of $c(-, \mathbb{A}^1)$ onto the constant sheaf $\mathbb{Z}$ extends to induce an $\mathbb{A}^1$-deformation of $c(-; \mathbb{A}^1)\tilde{\otimes}\mathcal{Z}$ onto $\mathcal{Z}$. Thus, using the factorization $\mathcal{Z} \subset c(-; \mathbb{A}^1)\tilde{\otimes}\mathcal{Z} \subset \mathcal{Z}_{tr}$ of the inclusion $\mathcal{Z} \subset \mathcal{Z}_{tr}$, we see that it suffices to prove that the last inclusion $c(-; \mathbb{A}^1)\tilde{\otimes}\mathcal{Z} \subset \mathcal{Z}_{tr}$ is an $\mathbb{A}^1$-equivalence, or in other words that the cokernel $\overline{\mathcal{Z}}_{tr}$ of the latter inclusion is $\mathbb{A}^1$-equivalent to 0.

To prove that $\overline{\mathcal{Z}}_{tr}$ is $\mathbb{A}^1$-trivial we will embedd it into an other sheaf of abelian groups $\overline{\mathcal{Z}}$ and we will show that both $\overline{\mathcal{Z}}_{tr}$ and the cokernel $\hat{\mathcal{Z}} := \overline{\mathcal{Z}}/\overline{\mathcal{Z}}_{tr}$ admit an explicit $\mathbb{A}^1$-homotopy to zero.

Now the cokernel $\overline{\mathcal{Z}}_{tr}$ has the following values on the henselization A of a point in a smooth k-scheme

$$\overline{\mathcal{Z}}_{tr}(A) = \oplus_{y \in \mathcal{C}(Spec(A), \mathbb{A}^1)} (\mathcal{Z}(\overline{y})/\mathcal{Z}(A))$$

where $\mathcal{Z}(\overline{y})/\mathcal{Z}(A)$ is the cokernel of the injective morphism

$$\mathcal{Z}(A) \subset \mathcal{Z}(\overline{y})$$

The sheaf $\overline{\mathcal{Z}}$ will have values on $Spec(A)$ the group

$$\overline{\mathcal{Z}}(A) = \oplus_{x \in (\mathbb{A}^1_F)^{(1)}} (\mathcal{Z}(A[x])/\mathcal{Z}(A))$$

where F is the fraction field of A, and for each closed point x of $\mathbb{A}^1_F$ we let $A[x] \subset \kappa(x)$ be the sub-A-algebra of the residue field $\kappa(x)$ of x generated by x itself, viewed as a function $Spec(\kappa(x)) \to \mathbb{A}^1_F$. We will write in the sequel $\mathcal{Z}_0(A[x])$ for $\mathcal{Z}(A[x])/\mathcal{Z}(A)$.

Observe that though $\kappa(x)$ is a finite extension of F, in general x is not integral over A and $A[x]$ is not a finite type $\mathbb{A}$-module. The x which have this property are exactly the one integral over A which corresponds exactly to the $y \in \mathcal{C}(Spec(A), \mathbb{A}^1)$ and gives the inclusion

$$\overline{\mathcal{Z}}_{tr} \subset \overline{\mathcal{Z}}$$

The only problem to construct $\overline{\mathcal{Z}}$ as a sheaf is to define the reduction morphisms

$$\overline{\mathcal{Z}}(A) \to \overline{\mathcal{Z}}(B)$$

where $A \twoheadrightarrow B$ is the morphism corresponding to a closed essentially smooth k-scubscheme $Spec(B) \subset Spec(A)$. The basic idea is simple. Take an x as above. The kernel of the epimorphism $A[X] \twoheadrightarrow A[x], X \mapsto x$ is of course generated by a primitive Polynomial $P_0 = a_n X^n + a_{n-1} X^{n-1} + \cdots + a_0 \in A[X]$ such that $\frac{1}{a_n} P_0 \in F[X]$ is the minimal polynomial P of x over F. Given an epimorphism $A \twoheadrightarrow B$ as above, one wants to define the reduction morphism on the summand $\mathcal{Z}_0(A[x])$ by using the epimorphism from the algebra $A[X]/(P_0) \cong A[x]$ onto $B[X]/(\overline{P}_0)$, where $\overline{P}_0 \in B[X]$ is the reduction of P_0. We proceed as follows. If $\overline{P}_0 \neq 0 \in B[X]$, as B is also a UFD, one may write

$$\overline{P}_0 = c \Pi_i Q_i^{n_i}$$

for some primitive irreducible polynomials Q_i of $B[X]$, and $c \in B - \{0\}$. We write x_i for the closed point in $\mathbb{A}^1_K$ corresponding to Q_i, where K is the fraction field of B. For each i we have a canonical epimorphism of A-algebras

$$A[x] \twoheadrightarrow B[x_i]$$

We then define the reduction morphism on the summand $\mathcal{Z}_0(A[x])$ as the sum

$$\Sigma_i [n_i]_\epsilon \mathcal{Z}_0(A[x] \twoheadrightarrow B[x_i]) : \mathcal{Z}_0(A[x]) \to \oplus_i \mathcal{Z}_0(B[x_i])$$

If $\overline{P}_0 = 0 \in B[X]$, we just take 0 for the reduction morphism on the summand $\mathcal{Z}_0(A[x])$.

LEMMA 5.4. *The previous definition can be extended in a unique way to a sheaf of abelian groups in the Nisnevich topology* $(Sm_k)^{op} \ni X \mapsto \overline{\mathcal{Z}}(X)$.

The proof is straightwforward. This construction works only because we work with the sum of the $\mathcal{Z}_0(A[x])$. There is no way to define a sheaf structure whose values on A would be $\oplus_{x\in(\mathbb{A}^1_F)^{(1)}}\mathcal{Z}(A[x])$. The main obstruction is when the reduction of P_0 in $B[X]$ vanishes (see above) one doesn't know where to send $\mathcal{Z}(A[x])$. Killing the $\mathcal{Z}(A)$'s everywhere kills precisely this obstruction; for the details see [**43**]. □

Now it remains to explicit on both $\overline{\mathcal{Z}}$ and the quotient $\hat{\mathcal{Z}} = \overline{\mathcal{Z}}/\overline{\mathcal{Z}}_{tr}$ an $\mathbb{A}^1$-homotopy to zero. For the first one we proceed as follows. For F a field extension of k, we have the obvious action of F on $\overline{\mathcal{Z}}(F) = \oplus_{x\in(\mathbb{A}^1_F)^{(1)}}\mathcal{Z}_0(\kappa(x))$ which let 0 acts by the zero morphism and which let $f \in F^\times$ acts by permuting the x's by $x \mapsto \frac{x}{f}$.

Lemma 5.5. *The previous construction extends uniquely to a morphism of sheaves $\mathbb{A}^1 \times \overline{\mathcal{Z}} \to \overline{\mathcal{Z}}$ which defines an $\mathbb{A}^1$-homotopy between the Identity morphism and the zero morphism.*

Proof. One has basically to check that given an henselization A of a point in a smooth k-scheme the multiplication by $f : \overline{\mathcal{Z}}(F) \to \overline{\mathcal{Z}}(F)$ preserves $\overline{\mathcal{Z}}(A)$ and induces $f : \overline{\mathcal{Z}}(A) \to \overline{\mathcal{Z}}(A)$ and that moreover given an epimorphism $A \twoheadrightarrow B$ as above, that the obvious diagram

$$\begin{array}{ccc} \overline{\mathcal{Z}}(A) & \xrightarrow{f} & \overline{\mathcal{Z}}(A) \\ \downarrow & & \downarrow \\ \overline{\mathcal{Z}}(B) & \xrightarrow{f} & \overline{\mathcal{Z}}(B) \end{array}$$

commutes. To check this one may reduce to the case $Spec(B) \subset Spec(A)$ is of codimension one, and then by localization also to the case B is a field. This now can be hand checked: if f is a unit it is clear, if not we may reduce to the case where f is an irreducible element of A, that is to say when f generates the maximal ideal of A. But then given a point x, in the sequence $f^i.x$ of points (related by the multiplication by f), there exists one and only one $i \in \mathbb{Z}$ such that $f^i.x \in \kappa(x)$ is integral over A and $f^{i-1}.x$ is not. In that case each summand $\mathcal{Z}_0(A[f^j.x])$, $j \neq i$, will reduce to 0 in $\mathcal{Z}_0(B)$ (as the point $f^j.x$ specializes to 0 for $j > i$ and to ∞ for $j < i$). This implies clearly the claim as both composition morphisms in the diagram are then zero. □

We know claim that the sheaf $\hat{\mathcal{Z}}$ is aslo $\mathbb{A}^1$-contractible by an explicit $\mathbb{A}^1$-homotopy to 0. More precisely for A the henselization of a point in a smooth k-scheme, with field of fractions F and residue field κ, we have

$$\hat{\mathcal{Z}}(A) = \oplus_{x\in(\mathbb{A}^1_F)^{(1)}|x_\infty\neq\emptyset}\mathcal{Z}_0(A(x))$$

where we mean $x_\infty \neq \emptyset$ if and only if $\overline{P}_0$ is constant in $\kappa[X]$ (observe that this holds exactly when the closure $\overline{x} \subset \mathbb{P}^1_A$ of the point x meets the section at infinity $Spec(A) \subset \mathbb{P}^1_A$).

We take the action of A on $\hat{\mathcal{Z}}(A)$ which let 0 acts by the zero morphism and which let $f \in A - \{0\}$ acts by mapping (again) x to $\frac{x}{f}$. This defines an $\mathbb{A}^1$-homotopy of the identity of $\hat{\mathcal{Z}}$ to 0 and proves the claim. □

REMARK 5.6. Contrary to the appearances, this homotopy, though very close to the previous one, is not compatible with the one on $\overline{\mathcal{Z}}$, and the epimorphism $\overline{\mathcal{Z}} \twoheadrightarrow \hat{\mathcal{Z}}$ doesn't preserve the $\mathbb{A}^1$-homotopy. In fact most probably, there is no $\mathbb{A}^1$-homotopy of the identity morphism of $\overline{\mathcal{Z}}_{tr}$ to 0. □

5.2. Transfers structures on the $H_*^{\mathbb{A}^1}((\mathbb{G}_m)^n)$'s and rigidity. We now write down explicitely the structures that we get on the homology sheaves $H_*^{\mathbb{A}^1}((\mathbb{G}_m)^n)$'s by the construction of the previous section.

First, the $\mathbb{Z}[\mathbb{G}_m]$-module structure can be seen to induce a structure of bGW-modules on these. Recall that $\underline{\mathbf{GW}}$ denotes the sheaf of unramified Grothendieck-Witt groups on Sm_k [**41**]. This is the strictly $\mathbb{A}^1$-invariant sheaf of unramified Milnor-Witt K-theory $\underline{\mathbf{K}}_0^{MW}$ in weight 0 constructed in [**41**]. This is a sheaf of commutative rings and its sections on a field $F \in \mathcal{F}_k$ is the ring of non-degenerated symmetric bilinear forms over F defined and studied in [**36**]. We denote as usual by $< a >\in GW(F)$ the form of rank 1 associated to a unit $a \in F^\times$. There is an induced rank homomorphism $\underline{\mathbf{GW}} \to \mathbb{Z}$. If M is a sheaf of $\underline{\mathbf{GW}}$-modules and $a \in F^\times$ a unit, $< a >: M(F) \to M(F), \mu \mapsto < a > \mu$ is thus the multiplication by the class $< a >\in GW(F)$.

For any integer $n \in \mathbb{N}$, we set

$$[n]_\epsilon = \sum_{i=0}^{n-1} < (-1)^i >\in GW(\kappa)$$

If $n = 2m$ is even then $[n]_\epsilon = m.h$ with $h = 1+ < -1 >$ being the hyperbolic plane. If $n = 2m + 1$ is odd, then $[n]_\epsilon = 1 + m.h$.

We will need the following notion[7] in the next definition. Given a field K of characteristic p and a unitary irreducible polynomial $P \in K[X]$ we may write canonically P as $P_0(X^{p^m})$ where P_0 is a unitary irreducible separable polynomial and $m \in \mathbb{N}$ an integer. This corresponds to the canonical factorization of $K \subset L = k[X]/P$ as $K \subset L_{sep} \subset L$ where $K \subset L_{sep}$ is separable and $L_{sep} \subset L$ purely inseparable. m is thus the smallest integer with $x^{p^m} \in L_{sep}$ and $[L : L_{sep}] = p^m$. P_0 is the minimal polynomial of x^{p^m} over K. We will denote by P_0' the derivative of P_0; thus $P_0' \neq 0$.

If $char(K) = 0$ then we just take $P = P_0$ and $m = 0$. Observe that if one has a root x of P in an extension $K \subset L$ then $P_0'(x) \in L^\times$ in any characteristic.

DEFINITION 5.7. Let M be a sheaf of $\underline{\mathbf{GW}}$-modules. A structure of *generalized transfers* on M consists of the following Datum:

[7] The reader who doesn't want to care about inseparability problem in finite characteristic might ignore these and asume everywhere that the fields (or rings) are perfect

(D) For each finite extension $\phi : K \subset L$ in $\mathcal{F}_k$ a morphism:

$$Tr(\phi) = Tr_K^L : M(L) \to M(K)$$

called the transfer morphism from L to K.

This Datum should satisfy the following 4 axioms:

(A0) Functoriality. $Tr_K^K = Id_{M(K)} : M(K) \to M(K)$ and for any finite extensions $K \subset F \subset L$ in $\mathcal{F}_k$ then one has $Tr_K^F \circ Tr_F^L = Tr_K^L : M(L) \to M(K)$.

(A1) Projection Formulas. Let $\phi : K \subset L$ in $\mathcal{F}_k$ be a finite extension. For $a \in K^\times$ and $\mu \in M(L)$ one has $Tr_K^L(< \phi(a) > \mu) =< a > Tr_K^L(\mu) \in M(K)$. In the same way, if $a \in L^\times$ and $\mu \in M(F)$ then one has $Tr_K^L(< a > .\phi(\mu)) = Tr_K^L(< a >).\mu \in M(K)$, where $Tr_K^L(< a >) \in GW(K)$ is the usual transfer from $< a >$.

(A2) Let $\phi : K \subset L$ be a finite extension in $\mathcal{F}_k$ and $\psi : K \subset E$ be an extension in $\mathcal{F}_k$. Let R be the tensor product $E \otimes_K L$, a finite E-algebra. For $x \in Spec(R)$ let $\phi_x : E \subset \kappa(x)$ and $\psi_x : L \subset \kappa(x)$ be the corresponding extensions in $\mathcal{F}_k$, with $\phi_x : E \subset \kappa(x)$ finite. Let λ_x be the length of the local ring R_x. Then

$$M(\psi) \circ Tr_K^L = \sum_{x \in Spec(R)} [\lambda_x]_\epsilon \; Tr_E^{\kappa(x)} \circ M(\psi_x)$$

(A3) Let $\phi : K \subset L$ be a finite extension in $\mathcal{F}_k$, let $A \subset K$ be a geometric discrete valuation ring of K with residue field κ. We assume that the integral closure B of A in L is also a discrete valuation ring with residue field λ. Let $\kappa \subset \lambda$ be the induced residue fields extension and let $e \in \mathbb{N}$ be the ramification index. We let $s_B : M(B) \to M(\lambda)$ and $s_A : M(A) \to M(\kappa)$ be the restriction maps.

(A3)(1) If $e = 1$ that is to say $[\lambda : \kappa] = [L : K]$, then Tr_K^L maps $M(B) \subset M(L)$ into $M(A) \subset M(K)$. Denote by $Tr_A^B : M(B) \to M(A)$ the induced morphism. Then moreover the following diagram commutes

$$\begin{array}{ccc} M(B) & \stackrel{s_B}{\to} & M(\lambda)) \\ Tr_A^B \downarrow & & \downarrow Tr_\kappa^\lambda \\ M(A) & \stackrel{s_A}{\to} & M(\kappa) \end{array}$$

(A3)(2) If $Spec(B) \to Spec(A)$ is monogenous, let $x \in B$ be a generator of B over A and let P be its minimal polynomial. Let $Q \in \kappa[X]$ be the minimal polynomial of the image $\overline{x}$ of x in λ; observe that $\overline{x}$ is a generator of λ over κ and that the mod κ reduction $\overline{P}$ of P is equal to Q^e. Then the composition $M(L) \stackrel{<P_0'(x)>}{\longrightarrow} M(L) \stackrel{Tr_K^L}{\to} M(K)$ maps $M(B)$ in $M(A)$. Let $\tau_A^B(x) : M(B) \to M(A)$ be the induced morphism and denote $\tau_\kappa^\lambda(\overline{x})$ the composition $M(\lambda) \stackrel{<Q_0'(x)>}{\longrightarrow} M(\lambda) \stackrel{Tr_K^L}{\to} M(\kappa)$. Then moreover the following diagram commutes

$$\begin{array}{ccc} M(B) & \stackrel{s_B}{\to} & M(\lambda) \\ \tau_A^B(x)\downarrow & & \downarrow [e]_\epsilon\, \tau_\kappa^\lambda(\overline{x}) \\ M(A) & \stackrel{s_A}{\to} & M(\kappa) \end{array}$$

EXAMPLE 5.8. For a given integer $n \in \mathbb{Z}$ the sheaves $\underline{\mathbf{K}}_n^{MW}$ may be shown to have such a structure, in a canonical way. The transfers morphisms are compatible with the one in Milnor K-theory through the projection $\underline{\mathbf{K}}_n^{MW} \to \underline{\mathbf{K}}_n^{M}$ and to the canonical transfers in $\underline{\mathbf{I}}^n$ (see [**53**]) through the projection $\underline{\mathbf{K}}_n^{MW} \to \underline{\mathbf{I}}^n$. □

REMARK 5.9. An $\mathbb{A}^1$-invariant sheaf with transfers M in the sense of Voevodsky defines such a structure of generalized transfers, in which the $\underline{\mathbf{GW}}$-module structure in Data (D2) is taken to be the trivial one, that is to say $\underline{\mathbf{GW}}$ acts through the rank $\underline{\mathbf{GW}} \to \mathbb{Z}$, and in which the Data (D1) comes by evaluating the transfers on the fields. We will see below in Theorem 5.16 that the converse holds. □

REMARK 5.10. For any $\mathbb{P}^1$ (or T) spectrum E the homotopy sheaves $\pi_n^{\mathbb{A}^1}(E)$ do have this structure. In fact a more careful (and non trivial) analysis shows that for any strictly $\mathbb{A}^1$-invariant sheaf M the sheaf M_{-1} admits a canonical structure of generalized transfers. It is still an open question to kow if the converse holds. Is it true that any strictly $\mathbb{A}^1$-invariant sheaf with generalized transfers is of the form M_{-1}? □

REMARK 5.11. From Axiom **(A3)(1)** it follows that in particular any finite étale morphism $Y \to X$ between smooth k-schemes the morphism $Tr_K^L : M(L) \to M(K)$ induces a canonical transfer morphism $M(B) \to M(A)$.

Note that the two parts of Axiom **(A3)** are not independant. For instance **(A3)(2)** in the case $e = 1$ implies **(A3)(1)** for a monogenous extension $A \subset B$. An other particular case of Axiom **(A3)(2)** is when $Spec(B) \to Spec(A)$ is totally ramified that is to say $e = [L : K]$ and $\kappa \subset \lambda$ is an isomorphism.

In characteristic 0 the two previous cases ($e = 1$ and $e = [L : K]$) are sufficient to characterize the generalized transfer structure. In finite characteristic it is a bit more tricky.

For a more conceptual approach on these Axioms involving modules of Kähler differential forms see [**41, 43**]. We gave the previous list of axioms, which doesn't seem "conceptual" because it is exactly the way we are going to check the properties by hands on the $\mathbb{A}^1$-complexes of tori in [**43**]. It would be interesting to have a more conceptual proof of these facts. □

Here is for us the most interesting example:

THEOREM 5.12. *Let n be an integer. Let k be a perfect field. The $\mathbb{Z}[\mathbb{G}_m]$-module structure and the transfers morphisms induced on the sheaves*

$$H_*^{\mathbb{A}^1}((\mathbb{G}_m)^{\wedge n}) = H_*(L_*^{\mathbb{A}^1}(\mathbb{Z}((\mathbb{G}_m)^{\wedge n})$$

by the constructions of the previous section define a structure of strictly $\mathbb{A}^1$-invariant sheaves with generalized transfers in the previous sense.

The proof is straightforward though quite lenghty [**43**]. □

The rigidity Theorem. We now state and prove a rigidity theorem for strictly $\mathbb{A}^1$-invariant sheaves with generalized transfers. The proof reduces to the "classical" rigidity Theorem of [**63**].

THEOREM 5.13. *Let M be a strictly $\mathbb{A}^1$-invariant sheaf of abelian groups with generalized transfers. Then M satisfies the rigidity property at ℓ.*

PROOF OF THEOREM 3. From Theorem 5.12 and the previous Theorem, applied to the field k being a prime field, we get Theorem 4.10 which we already know implies Theorem 3 as well. □

Let us mention a stronger form of Theorem 5.13 which is not really needed as far as the Friedlander-Milnor conjecture is concerned:

THEOREM 5.14. *Assume that if $\ell = 2$ then any field $F \in \mathcal{F}_k$ is of finite virtual cohomological 2-dimension. Let M be a strictly $\mathbb{A}^1$-invariant sheaf of abelian groups with generalized transfers. Then M satisfies the strong rigidity property at ℓ.*

For the case $\ell \neq 2$, in fact the proof of 5.13 follows directly from the classical rigidity theorem [**63**] and the following observation (observe that to prove Theorem 5.13 we may always assume that -1 is a square):

LEMMA 5.15. *Assume -1 is a square in k. Let M be a strictly $\mathbb{A}^1$-invariant sheaf of $\mathbb{Z}[\frac{1}{2}]$-modules with generalized transfers. Then the $\underline{\mathbf{GW}}$-module structure is trivial and thus M is in fact a sheaf with transfers in the sense of Voevodsky by Theorem 5.16 below.*

PROOF. By Theorem 5.16 it suffices to prove that the $\underline{\mathbf{GW}}$-module structure is trivial. But as -1 is a square, $1+ < -1 >= 2$ is 0 in $I(k)$. The sheaf $\underline{\mathbf{I}}$ is thus a sheaf of $\mathbb{Z}/2$-modules, which automatically acts trivially on a $\mathbb{Z}[\frac{1}{2}]$-module. Thus the $\underline{\mathbf{GW}}$-module structure is indeed trivial. □

The following establishes a converse to Remark 5.9.

THEOREM 5.16. *Let M be a strictly $\mathbb{A}^1$-invariant sheaf of abelian groups with a structure of generalized transfers. Assume that the structure of $\underline{\mathbf{GW}}$-modules is trivial in the previous sense. Then there exists a unique structure of sheaf with transfers on M in the sense of Voevodsky which induces the structure of generalized transfers.*

To treat the case $\ell = 2$ (as well as the more general Theorem 5.14) we will have to work a bit more. More precisely we observe:

THEOREM 5.17. [**43**] *Let M be a strictly $\mathbb{A}^1$-invariant sheaf M with generalized transfers. There exists a unique sub-sheaf $\mathcal{I}M \subset M$ which has the following properties:*

1) *the group of sections $\mathcal{I}M(F)$ on a field $F \in \mathcal{F}_k$ is the sub-$GW(F)$-module generated by the images $Tr_K^L(I(L).M(L))$ of the $I(L).M(L)$ through the transfer morphism $Tr_K^L : M(L) \to M(K)$ where L runs over each finite extensions of K;*
2) *$\mathcal{I}M$ is strictly $\mathbb{A}^1$-invariant;*

Moreover, $\mathcal{I}M$ is stable both by the structure of $\underline{\mathbf{GW}}$-module and the structure of generalized transfers.

REMARK 5.18. In general, it is not true that for a field $\mathcal{I}M(F)$ is just $I(F).M(F)$. This is true in the case of the sheaves of the form $\underline{\mathbf{I}}^n$ for instance by one of the results of [**1**]. □

From what we have seen up to now we get:

COROLLARY 5.19. *Let M be a strictly $\mathbb{A}^1$-invariant sheaf M with generalized transfers. Then the quotient sheaf*

$$M/\mathcal{I}M$$

is in a canonical way a sheaf with transfers in the sense of Voevodsky, and as a consequence has the strong rigidity property.

PROOF. By the previous Theorem $M/\mathcal{I}M$ is a strictly $\mathbb{A}^1$-invariant sheaf with generalized transfers in which the $\underline{\mathbf{GW}}$-module structure is trivial. The result follows now from Theorem 5.16. □

PROOF OF THEOREM 5.13. Let M be a strictly $\mathbb{A}^1$-invariant sheaf M with generalized transfers. We may assume that $\ell = 2$ and that M is a sheaf of $\mathbb{Z}/2$-modules (by Lemma 5.15). Let A be the (strict) henselization of a geometric point x in a smooth k-scheme X.

We claim that the filtration $\ldots \mathcal{I}^n M(A) \subset \mathcal{I}^{n-1}M(A) \ldots$ is finite. This follows from the previous Theorem, the affirmation by Voevodsky of the Milnor conjectures which implies (see [**2**, Lemma 2.1]) that for a field F of finite type over its prime field, $I^{n+1}(F) = 2I^n(F)$ for n large enough.

Thus to prove that the rigidty property it suffices to check that each of the morphism $I^n M/I^{n+1}M(A) \to I^n M/I^{n+1}M(\kappa)$ is an isomorphism, where κ is residue field. But this follows from the previous corollary and the classical rigidity property [**63**]. □

A $\mathbb{A}^1$-coverings and Friedlander conjecture in degree 2

In this section we give a proof of the Friedlander conjecture in degree 2.

A.1. $\mathbb{A}^1$-coverings and central extensions of algebraic groups. Recall that G denotes a split semi-simple k-group. We want to study a bit the sheaf $\pi_0^{\mathbb{A}^1}(G)$ defined in [**44**]. It is clearly a sheaf of groups in the Nisnevich topology.

If moreover G is assumed to be simply connected then for each field extension F/k the group $G(F)$ is generated by the images of the homomorphisms $\mathbb{G}_a \to G$ over F, where $\mathbb{G}_a$ is the additive group see [**65**, §1.1.2]. Consequently G is weakly 0-connected and thus $\mathbb{A}^1$-connected by [**38**, Lemma 6.1.3], which means that:

LEMMA A.1. *Let G be a split simply connected semi-simple k-group, then*

$$\pi_0^{\mathbb{A}^1}(G) = *$$

Now we will use this result to study the $\pi_0^{\mathbb{A}^1}(G)$ for G an arbitrary split semi-simple k-group. Let $T \subset G$ be a split maximal Torus. Let $\tilde{G} \to G$ be the universal covering in the sense of algebraic groups theory. Let $C \subset \tilde{G}$ be the kernel; it is a finite central abelian group and the morphism a Galois étale covering with Galois group C. Let $\tilde{T} \subset \tilde{G}$ be the inverse image of T. It is a split maximal torus. Let us denote by $T/_{Nis}\tilde{T}$ the quotient sheaf in the Nisnevich topology. This is an $\mathbb{A}^1$-invariant sheaf with transfers in the sense of Voevodsky. The étale short exact sequence $0 \to C \to \tilde{T} \to T \to 0$ and Hilbert Theorem 90 ($\tilde{T}$ being also split) gives an isomorphism of sheaves in the Nisnevich topology

$$T/_{Nis}\tilde{T} \to \mathcal{H}^1_{et}(C)$$

where $\mathcal{H}^1_{et}(C)$ is the associated sheaf to $X \mapsto H^1_{et}(X; C)$.

THEOREM A.2. *Keeping the above notations and assumptions, the morphism of sheaves $T \to \pi_0^{\mathbb{A}^1}(G)$ induced by the inclusion $T \subset G$ is onto and induces an isomorphism*

$$T/_{Nis}\tilde{T} \cong \pi_0^{\mathbb{A}^1}(G)$$

Moreover this sheaf clearly vanishes in the étale topology.

PROOF. It follows from the following commutative diagram of smooth k-schemes

$$\begin{array}{ccc} C & = & C \\ \downarrow & & \downarrow \\ \tilde{T} & \subset & \tilde{G} \\ \downarrow & & \downarrow \\ T & \subset & G \end{array}$$

whose bottom square is cartesian. First it induces a isomorphism of quotient smooth k-schemes $\tilde{G}/\tilde{T} \cong G/T$. Now by the previous Theorem, $\tilde{G}$ being simply connected (in the sense of algebraic group theory) is $\mathbb{A}^1$-connected. But the morphism $\tilde{G} \to \tilde{G}/\tilde{T}$ is Zariski (thus Nisnevich) locally trivial and induces an isomorphism on the $\pi_0^{\mathbb{A}^1}$. Thus G/T is $\mathbb{A}^1$-connected.

As $\tilde{G} \to \tilde{G}/\tilde{T}$ and $G \to G/T$ are $\mathbb{A}^1$-coverings (being a Zariski locally trivial T-torsor, with T split) [**41**] we get long exact sequences of $\mathbb{A}^1$-homotopy sheaves.

The one for $G \to G/T$ gives an exact sequence (in the Nisnevich topology) of the form

$$\pi_1^{\mathbb{A}^1}(G/T) \to T \to \pi_0^{\mathbb{A}^1}(G) \to 0$$

and the second one gives that $\pi_1^{\mathbb{A}^1}(G/T) \cong \pi_1^{\mathbb{A}^1}(\tilde{G}/\tilde{T}) \to \tilde{T} \to \pi_0^{\mathbb{A}^1}(\tilde{G}) = 0$ is onto. This clearly implies the claim. □

REMARK A.3. In the case $G = SO'_n$ (split) for instance the quotient sheaf $T/_{Nis}T'$ is isomorphic to the sheaf $\mathbb{G}_m/(\mathbb{G}_m)^2$ and the morphism $SO'_n \to \pi_0^{\mathbb{A}^1}(SO'_n) = \mathbb{G}_m/(\mathbb{G}_m)^2$ is the spinor norm. □

Let $G^{(0)} \subset G$ be the kernel of $G \to \pi_0^{\mathbb{A}^1}(G)$.

LEMMA A.4. *The induced morphism $\tilde{G} \to G^{(0)}$ is an epimorphism in the Nisnevich topology.*

PROOF. One use the short exact sequence in the étale topology $0 \to C \to \tilde{G} \to G$ to get for any smooth F-scheme X an associated long exact of cohomology groups/sets

$$\tilde{G}(X) \to G(X) \to H^1_{et}(X; C) \to \cdots$$

by sheafifying in the Nisnevich topology we get the exact sequence:

$$\tilde{G} \to G \to \mathcal{H}^1_{et}(C)$$

which easily implie the claim taking the above isomorphism $\pi_0^{\mathbb{A}^1}(G) \cong \mathcal{H}^1_{et}(C)$ into account. □

By construction $G^{(0)}$ is an $\mathbb{A}^1$-connected space with a base point (the neutral element). Thus it admits by [**41**] a universal $\mathbb{A}^1$-covering $\tilde{G}^{\mathbb{A}^1} \to G^{(0)}$ which is easily seen as usual to define a central extension in the Nisnevich topology of the form

$$0 \to \pi_1^{\mathbb{A}^1}(G) \to \tilde{G}^{\mathbb{A}^1} \to G^{(0)} \to 1 \tag{A.1}$$

If G is simply connected, it follows from the previous Theorem that $\pi_0^{\mathbb{A}^1}(G) = *$ and that the previous extension gives a central extension in the Nisnevich topology[8] of the form

$$0 \to \pi_1^{\mathbb{A}^1}(G) \to \tilde{G}^{\mathbb{A}^1} \to G \to 1 \tag{A.2}$$

Now for G general (non necessarily simply connected) we see that the associated morphism of sheaves in the étale topology to $G^{(0)} \to G$ is an isomorphism as this follows from Theorem A.2 above. Consequently the

[8]actually in the Zariski topology as well

associated central extension to (A.1) in the étale topology has the following form

$$0 \to \pi_1^{\mathbb{A}^1}(G)_{et} \to \tilde{G}_{et}^{\mathbb{A}^1} \to G \to 1 \tag{A.3}$$

This extension has the following universal property:

THEOREM A.5. *Given a central extension in the étale topology*

$$0 \to M \to E \to G \to 1$$

such that M is a locally constant étale sheaf of torsion prime to $char(k)$, then there exists one and only one morphism of sheaves in the étale topology $\pi_1^{\mathbb{A}^1}(G)_{et} \to M$ which induces the given extension from (A.3). In other words:

$$H^2_{et}(G; M) \cong Hom_{\mathcal{A}b(k)}(\pi_1^{\mathbb{A}^1}(G), M)$$

where $\mathcal{A}b(k)$ means the category of sheaves of abelian groups in the Nisnevich topology on Sm_k.

PROOF. Given a central extension in the étale topology $0 \to M \to E \to G \to 1$ with M as above, the morphism $E \to G$ is an $\mathbb{A}^1$-covering [**41**]. From the universal property of the universal $\mathbb{A}^1$-covering, we get a morphism $\overline{G}^{\mathbb{A}^1} \to E$; the associated morphism in the étale topology has the required property. Uniqueness is easy to establish along the same lines. □

REMARK A.6. The previous Theorem can be strenghened as follows: say that a sheaf of abelian groups M is the étale topology is strictly étale $\mathbb{A}^1$-invariant if $X \mapsto H^i_{et}(X; M)$ is $\mathbb{A}^1$-invariant for any $i \in \mathbb{N}$. For instance any $\mathbb{A}^1$-invariant étale sheaf of $\mathbb{Z}[\frac{1}{p}]$-modules[9] with transfers in the sense of Voevodsky is strictly étale $\mathbb{A}^1$-invariant by [**68**, Corollary 5.29 p.136]. Then one may prove that $\pi_1^{\mathbb{A}^1}(G)[\frac{1}{p}]_{et}$ is srongly étale $\mathbb{A}^1$-invariant and that the "induced" central extension

$$0 \to \pi_1^{\mathbb{A}^1}(G)[\frac{1}{p}]_{et} \to \tilde{G}'_{et} \to G \to 1$$

is the universal central extension with kernel a strictly étale $\mathbb{A}^1$-invariant sheaf of $\mathbb{Z}[\frac{1}{p}]$-modules. □

A.2. Proof of the Friedlander conjecture in degree 2. In this section we give a proof of the Friedlander conjecture in degree 2 using the construction of the previous section. Using the classical devissage and reductions, we see that it suffices to check the case G is simple and simply connected (in the sense of algebraic group theory), and F is separably closed.

The left hand side of the Friedlander conjecture is the group $H^2_{et}(BG; \mathbb{Z}/\ell)$ that is to say the group of isomorphism classes of central extensions

$$0 \to \mathbb{Z}/\ell \to E \to G \to 1$$

[9] here p is the characteristic exponent of F

of sheaves of groups in the étale topology. From the Theorem A.5, this group is isomorphic to $Hom_{\mathcal{A}b(k)}(\pi_1^{\mathbb{A}^1}(G), \mathbb{Z}/\ell)$.

On the other hands, the right hand side is the group $H^2(BG(F);\mathbb{Z}/\ell)$ of isomorphism classes of central extensions

$$0 \to \mathbb{Z}/\ell \to E \to G(F) \to 1$$

The group $G(F)$ is known to be perfect, F being infinite, so it admits a universal central extension of the form[10]

$$0 \to H_2(BG(F)) \to \widetilde{G(F)}^{alg} \to G(F) \to 1$$

and $H^2(BG(F);\mathbb{Z}/\ell) \cong Hom(H_2(BG(F), \mathbb{Z}/\ell)$. We now claim the following, which establish the Friedlander conjecture in degree 2.

LEMMA A.7. *With G as above and F separably closed,*

$$Hom(H_2(BG(F), \mathbb{Z}/\ell) = Hom_{\mathcal{A}b(k)}(\pi_1^{\mathbb{A}^1}(G), \mathbb{Z}/\ell) = 0$$

PROOF. $Hom(H_2(BG(F)), \mathbb{Z}/\ell) = 0$ follows from the classical computations of Matsumoto [**33**] which shows that $H_2(BG(F)) \cong K_2^M(F)$ and thus is ℓ-divisible. The vanishing of $Hom_{\mathcal{A}b(k)}(\pi_1^{\mathbb{A}^1}(G), \mathbb{Z}/\ell)$ follows from the fact that the inclusion $SL_2 \to G$ coming from a root of maximal length induces an epimorphism of sheaves (in the Nisnevich topology):

$$\underline{\mathbf{K}}_2^{MW} \cong \pi_1^{\mathbb{A}^1}(SL_2) \twoheadrightarrow \pi_1^{\mathbb{A}^1}(G)$$

This can be done using for instance [**57**, Theorem 4.1]. □

REMARK A.8. It can be shown that for G a split simply connected absolutely simple F-group, F arbitrary, then the previous epimorphism $K_2^{MW} \cong \pi_1^{\mathbb{A}^1}(SL_2) \twoheadrightarrow \pi_1^{\mathbb{A}^1}(G)$ induces an isomorphism of sheaves

$$\underline{\mathbf{K}}_2^{M} \cong \pi_1^{\mathbb{A}^1}(G)$$

except when G is of symplectic type, in which case

$$\underline{\mathbf{K}}_2^{MW} \cong \pi_1^{\mathbb{A}^1}(G)$$

is an isomorphism. Also it can be shown that the central extension

$$0 \to \underline{\mathbf{K}}_2^{M} \to E \to G \to 1$$

induced by (A.2) and the epimorphism $K_2^{MW} \twoheadrightarrow \pi_1^{\mathbb{A}^1}(G)$ is the one constructed [**6**], at least for F infinite.

Evaluting the universal extension (A.2) on F and comparing with the previous one of $G(F)$, we get a canonical homomorphism $H_2(BG(F)) \to$

[10]to avoid notational contradictions, we choosed to write $H_2(BG(F))$ for the group which is often denoted by $H_2(G(F))$

$\pi_1^{\mathbb{A}^1}(G)(F)$. It is also possible to show in fact that it is always an isomorphism (except maybe for very small finite fields). In other words we get an isomorphism, for any infinite field F

$$H_2(BG(F)) \cong H_2^{\mathbb{A}^1}(BG)(F) = \pi_1^{\mathbb{A}^1}(G)(F)$$

This is a stronger result as the Friedlander conjecture in degree 2. □

References

[1] J. Arason, Cohomologische invarianten quadratischer Formen. (German) J. Algebra 36 (1975), no. 3, 448–491.

[2] J. Arason and R. Elman, Powers of the Fundamental Ideal in the Witt Ring, J. Algebra 239 (2001).

[3] M. Artin and B. Mazur, Étale Homotopy, Lect. Notes in Maths 100, Springer Heidelberg 1969.

[4] N. Bourbaki, Groupes et algèbres de Lie, Chapitre 1.

[5] A.K. Bousfield and D.M. Kan, Homotopy limits, completions and localizations. Lecture Notes in Mathematics, Vol. 304. Springer-Verlag, Berlin-New York, 1972.

[6] J.-L. Brylinski and P. Deligne, Central extensions of reductive groups by $\mathbf{K}_2$. Publ. Math. Inst. Hautes tudes Sci. No. 94 (2001), 5–85.

[7] Séminaire Henri Cartan, 7 no. 1, 1954-1955, Algbre d'Eilenberg-Maclane et homotopie.

[8] E. Curtis, Some relations between homotopy and homology, Ann. of Maths. **83** (1965), 159–171.

[9] E. Curtis, simplicial homotopy theory, Advances in Maths.

[10] A. Dold, Homology of Symmetric Products and Other Functors of Complexes, Annals of Math.. 68 (1958), 54–80.

[11] A. Dold, Les foncteurs dérivés d'un foncteur non-additif. Séminaire Bourbaki, 5 (1958–1960), Exposé No. 170.

[12] E. Friedlander, The Friedlander-Milnor conjecture, L'enseignement Mathématique, (2) 54, 2008, 90–92.

[13] E. Friedlander, Étale homotopy of simplicial schemes, Ann. of Math. Sudies, PUP.

[14] E. Friedlander and G. Mislin, Cohomology of classifying spaces of complex Lie groups and related discrete groups. Comment. Math. Helv. 59 (1984), no. 3, 347–361.

[15] O. Gabber, K-theory of Henselian local rings and Henselian pairs, Contemp. Math. 126 (1992), 59–70.

[16] O. Gabber, Gersten's conjecture for some complexes of vanishing cycles. Manuscripta Math. 85 (1994), no. 3–4, 323–343.

[17] S.M. Gersten, Higher K-theory of rings, LNM 341.

[18] H. Gillet and R. Thomason, The K-theory of strict hensel local rings and a theorem of Suslin, J. Pure Appl. Algebra **34** (1984), 241–254.

[19] A. Grothendieck and J. Dieudonné, Éléments de Géométrie Algébrique IV. Étude locale des schémas et des morphismes de schémas (Seconde Partie). (French) Inst. Hautes Études Sci. Publ. Math. No. 24, 1965.

[20] A. Grothendieck and J. Dieudonné, Éléments de Géométrie Algébrique IV. Étude locale des schémas et des morphismes de schémas (Troisième Partie). (French) Inst. Hautes Études Sci. Publ. Math. No. 28, 1967, 361 pp.

[21] A. Grothendieck and J. Dieudonné, Éléments de Géométrie Algébrique IV. Étude locale des schémas et des morphismes de schémas (Quatrième Partie). (French) Inst. Hautes Études Sci. Publ. Math. No. 32, 1967.

[22] A. Grothendieck, Séminaire de Géométrie Algébrique, Revêtéments étales et Groupe Fondamental, L.N.M. 224.

[23] A. Grothendieck, Séminaire de Géométrie Algébrique, Théorie des Topos et cohomologie étale des schémas (SGA 4), Lect. Notes in Maths 269,270, et 305, Springer Heidelberg, 1972–73.

[24] F. Grunewald, J. Mennicke and L. Vaserstein: On the groups $\mathrm{SL}_2(\mathbf{Z}[x])$ and $\mathrm{SL}_2(k[x,y])$. Israel J. Math. 86 (1994), no. 1–3, 157–193.

[25] R. Hartschorne, Algebraic geometry, Springer 1977.

[26] M. Hovey, Model category structures on chain complexes of sheaves, Transactions of the AMS 353 (2001), 2441–2457.

[27] J. F. Jardine, On the homotopy groups of algebraic groups. J. Algebra 81 (1983), no. 1, 180–201.

[28] J. F. Jardine, Simplicial objects in a Grothendieck topos, Contemp. Math. 55 (1986), 193–239.

[29] M. Karoubi, Relations between algebraic K-theory and Hermitian K-theory. Journal of Pure and Applied Algebra 34, p. 259–263 (1984).

[30] K. Knudson, Homology of linear groups, Birkhäuser, Basel-Boston-Berlin.

[31] K. Knudson, The homology of special linear groups over polynomial rings, Ann. Sci. École Norm. Sup. (4) 30 (1997), no 3, 385–415.

[32] T.Y. Lam, Serre's conjecture, L.N.M. 635.

[33] H. Matsumoto, Sur les sous-groupes arithmétiques des groupes semi-simples déployés. (French) Ann. Sci. École Norm. Sup. (4) 2 1969, 1–62.

[34] J. Milne, Étale cohomology, Princeton University Press.

[35] J. Milnor, On the homology of Lie groups made discrete, Comment. Math. Helvetici 58 (1953) 072–085.

[36] J. Milnor, D. Husemoller, *Symmetric bilinear forms*, Ergebnisse der Mathematik, Band 73, Springer-Verlag.

[37] F. Morel, Sur les puissances de l'idéal fondamental de l'anneau de Witt, Commentarii Mathematici Helvetici, Vol 79 no 4, (2004) p 689.

[38] F. Morel, The stable $\mathbb{A}^1$-connectivity theorems, K-theory (2005) 35:p.1–68.

[39] F. Morel, $\mathbb{A}^1$-algebraic topology, Proceedings of the International Congress of Mathematicians, Madrid 2006.

[40] F. Morel, Milnor's conjecture on quadratic forms and mod 2 motivic complexes, Rend. Sem. Mat. Univ. Padova, Vol 114 (2005) pp 63–101.

[41] F. Morel, $\mathbb{A}^1$-algebraic topology over a field, preprint (2011).

[42] F. Morel, The weak Friedlander-Milnor conjecture, preprint (2010).

[43] F. Morel, Generalized transfers and the rigidity property for the $C_*^{\mathbb{A}^1}((\mathbb{G}_m)^{\wedge n})$'s, in preparation.

[44] F. Morel and V. Voevodsky, $\mathbb{A}^1$-homotopy theory of schemes, Publications Mathématiques de l'I.H.É.S. no 90.

[45] L.-F. Moser, $\mathbb{A}^1$-locality results for linear algebraic groups, in preparation.

[46] M. Pavaman Murthy, Zero cycles and projective modules, Ann. of Maths, 140 (1994).

[47] Y. Nisnevich, The completely decomposed topology on schemes and associated descent spectral sequences in algebraic K-theory. Algebraic K-theory: connections with geometry and topology (Lake Louise, AB, 1987), 241–342, NATO Adv. Sci. Inst. Ser. C Math. Phys. Sci., 279, Kluwer Acad. Publ., Dordrecht, 1989.

[48] D. Orlov, A. Vishik and V. Voevodsky, An exact sequence for K* M/2 with applications to quadratic forms Annals of Mathematics, 165(1), 1–13. 2007.

[49] D. Quillen, Homotopical algebra, L.N.M.

[50] C-H. Sah, Homology of classical Lie groups made discrete. III. J. Pure Appl. Algebra 56 (1989), no. 3, 269–312.

[51] J.W. Schlesinger, The semi-simplicial free Lie ring, Trans. Amer. Math. Soc. 122 (1966), 436–442.

[52] W. Scharlau, Quadratic and Hermitian forms. Grundlehren der Mathematischen Wissenschaften [Fundamental Principles of Mathematical Sciences], 270. Springer-Verlag, Berlin, 1985.
[53] Schmid, Wittring Homology, Phd Dissertation, Universitaet Regensburg, 1998.
[54] J.-P. Serre, Corps locaux, Hermann, Paris 1968.
[55] T. Springer, Linear algebraic groups. Progress in Mathematics, Volume 9, Birkhäuser.
[56] Soulé, Christophe Chevalley groups over polynomial rings. Homological group theory (Proc. Sympos., Durham, 1977), pp. 359–367, London Math. Soc. Lecture Note Ser., 36, Cambridge Univ. Press, Cambridge-New York, 1979.
[57] M. Stein, Chevalley groups over commutative rings. Bull. Amer. Math. Soc. 77 (1971), 247–252.
[58] D. Sullivan, Genetics of homotopy theory and the Adams conjecture, Ann. of Math, 100.
[59] A. Suslin, The structure of the special linear group over rings of polynomials. (Russian) Izv. Akad. Nauk SSSR Ser. Mat. 41 (1977), no. 2, 235–252, 477.
[60] A. Suslin, On the K-theory of algebraically closed fields, Invent. Math. 73 (1983), no. 2, 241–245.
[61] A. Suslin, On the K-theory of local fields. Proceedings of the Luminy conference on algebraic K-theory (Luminy, 1983). J. Pure Appl. Algebra 34 (1984), no. 2–3, 301–318.
[62] A. Suslin, Homology of GL_n, characteristic classes and Milnor K-theory. Algebraic K-theory, number theory, geometry and analysis (Bielefeld, 1982), 357–375, Lecture Notes in Math., 1046, Springer, Berlin, 1984.
[63] A. Suslin and V. Voevodsky, Singular homology of abstract algebraic varieties, Invent. Math. 123 (1996) 63 94.
[64] A. Suslin and V. Voevodsky, Relative cycles and Chow sheaves, Cycles, transfers, and motivic homology theories, 188–238, Ann. of Math. Stud., 143, Princeton Univ. Press, Princeton, NJ, 2000.
[65] J. Tits, Groupes de Whitehead de groupes algébriques simples sur un corps, Séminaire Bourbaki, Exposé No. 505, 1976/77.
[66] V. Voevodsky. Motivic Cohomology with Z/2-coefficients, Inst. Hautes Études Sci. Publ. Math., vol. 98 (2003), 59–104.
[67] V. Voevodsky, Triangulated categories of motives over a field. Cycles, transfers, and motivic homology theories, 188–238, Ann. of Math. Stud., 143, Princeton Univ. Press, Princeton, NJ, 2000.
[68] V. Voevodsky, Cohomological Theory of Presheaves with Transfers. Cycles, transfers, and motivic homology theories, 87–137, Ann. of Math. Stud., 143, Princeton Univ. Press, Princeton, NJ, 2000.
[69] M. Wendt, $\mathbb{A}^1$-homotopy of Chevalley groups, J. K-Theory 5 (2), 2010, pp. 245–287.
[70] M. Wendt, Homotopy Invariance for Homology of Linear Groups: Reductions and the rank two case. Preprint. Available at http://arxiv.org/abs/1011.5110.
[71] M. Wendt, Homotopy Invariance for Homology of Linear Groups: The Case SL4. Preprint. Available at http://arxiv.org/abs/1011.5392.

Current Developments in Mathematics, 2010

Universal formulas for counting Nodal curves on surfaces

Yu-Jong Tzeng

Abstract. These notes are supplementary to the author's lectures at the 2010 conference on Current Developments in Mathematics, held on November 19-20 in Cambridge Massachusetts.

1. Introduction

The aim of enumerative geometry is to find for the number of geometric objects of specified kind, which satisfy certain incidence or tangency constraints. The history of enumerative geometry can be traced back to Italian schools, where mathematicians raised many questions about algebraic varieties in $\mathbb{P}^N$ and obtained important formulas. In the last twenty years, techniques from superstring theory reinvigorated enumerative geometry, resulted in many exciting results such as Candelas et al.'s formula on genus zero Gromov-Witten invariants on Calabi-Yau quintics and the Yau-Zaslow formula on enumerating rational curves on K3 surfaces. Many classical questions were solved by introducing modern techniques. The most famous example is Kontsevich and Manin's recursive formula on N_d, the number of rational curves of degree d passing through $3d-1$ points in general position in $\mathbb{P}^2$.

In this article, we will report the recent progress of enumerating the number of nodal curves on algebraic surfaces.

1.1. Main results. Consider a line bundle L on a complex projective smooth surface S. This paper attempts to answer the following question: how many reduced curves have exactly r simple nodes and no higher singularities in a generic r-dimensional linear system of $|L|$? Equivalently, how many r-nodal curves in $|L|$ pass through $\dim|L| - r$ points in general position?

Göttsche [**Gö**] conjectured that for every r, the numbers of r-nodal curves are given by universal polynomials of four topological numbers: L^2, LK, $c_1(S)^2$ and $c_2(S)$ provided that the line bundle L is $(5r-1)$-very ample (i.e. $H^0(S,L) \to L|_\xi$ is surjective for every $\xi \in S^{[5r]}$). These polynomials are universal in the sense that it only depends on r and are independent of the surface and line bundle. Our first result is an algebro-geometric proof of Göttsche's conjecture.

THEOREM 1.1 (Göttsche's conjecture). *For every integer $r \geq 0$, there exists a universal polynomial $T_r(x,y,z,t)$ of degree r with the following property: given a smooth projective surface S and a $(5r-1)$-very ample (5-very ample if $r=1$) line bundle L on S, a general r-dimensional sublinear system of $|L|$ contains exactly $T_r(L^2, LK, c_1(S)^2, c_2(S))$ r-nodal curves.*

Since the numbers of nodal curves for all line bundles on $\mathbb{P}^2$ and for primitive classes on K3 surfaces have been determined ([**CH**], [**BL**]), all coefficients of T_r can be computed by solving linear equations. Moreover, one can combine all universal polynomials as coefficients to define a generating function. Inspired by the Yau-Zaslow formula, Göttsche [**Gö**] conjectured the closed form of this generating function of T_r, which we call the Göttsche-Yau-Zaslow formula.

THEOREM 1.2 (The Göttsche-Yau-Zaslow formula). *There exist universal power series $B_1(q)$ and $B_2(q)$ such that*

$$\sum_{r\geq 0} T_r(L^2, LK, c_1(S)^2, c_2(S))(DG_2(\tau))^r$$
$$= \frac{(DG_2(\tau)/q)^{\chi(L)} B_1(q)^{K_S^2} B_2(q)^{LK_S}}{(\Delta(\tau)D^2G_2(\tau)/q^2)^{\chi(\mathcal{O}_S)/2}},$$

where G_2 is the second Eisenstein series $-\frac{1}{24} + \sum_{n>0}\left(\sum_{d|n} d\right)q^n$, $D = q\frac{d}{dq}$ and $\Delta(q) = q\prod_{k>0}(1-q^k)^{24}$.

If we write $q = e^{2\pi i\tau}$ for τ on the complex upper half plane, then G_2, DG_2 and D^2G_2 are quasimodular forms in τ and Δ is a modular form in τ. For the precise definition of quasimodular forms, see [**KZ**].

1.2. Background. On $\mathbb{P}^2$, the number of nodal curves is classically known as the Severi degree $N_{d,g}$, which is the number of plane curves of degree d and genus g passing through $3d+g-1$ points in general position. This subject was studied by Ran ([**Ran1**], [**Ran2**]), using degeneration of $\mathbb{P}^2$ and an inductive procedure. In 1993, Kontsevich and Manin [**KM**] introduced the techniques of Gromov-Witten theory to this problem, from which they obtained a beautiful recursive formula of rational curves for all degrees. For plane curves of a higher genus, Harris and Pandharipande [**HP**] computed the Severi degrees with at most three nodes using Hilbert schemes, and Choi [**Ch**] extended the result to at most four nodes using Ran's method.

The counting of nodal curves of arbitrary genus g in $\mathbb{P}^2$ was completely solved by Caporaso and Harris [**CH**]. They defined the generalized Severi degrees $N_{d,g}(\alpha, \beta)$ with tangential conditions and used deformation theory to derive recursive formulas of $N_{d,g}(\alpha, \beta)$. Shortly after, Vakil [**Vakil**] derived similar results for rational ruled surfaces.

For an arbitrary smooth projective surface, the number of nodal curves with at most three nodes can be computed directly by standard intersection theory. In 1994, Vainsencher [**Va**] proved the existence of universal polynomials in the case of up to six nodes. By computing the polynomials explicitly, he showed that the polynomials only depend on L^2, LK, $c_1(S)^2$ and $c_2(S)$. Later, Kleiman and Piene ([**KP1**], [**KP2**]) refined Vainsencher's approach and generalized the result to up to eight nodes. Since their methods rely on a detailed analysis of the singularities of low codimensions, it is difficult to generalize the methods to the case of a higher number of nodes.

On algebraic K3 surfaces and primitive classes, the number of rational curves not only can be determined but also possesses a general pattern. Yau and Zaslow [**YZ**] discovered a surprising formula for the generating function in terms of the Dedekind function, which prompts the speculation that general modular forms may be involved. In particular, the Yau-Zaslow formula implies that the number of rational curves in an effective class C only depends on the self-intersection number C^2.

The Yau-Zaslow formula was generalized by Göttsche [**Gö**] to arbitrary projective surface. Using Vainsencher and Kleiman-Piene's numbers, Göttsche conjectured the Göttsche-Yau-Zaslow formula, which relates the generating function to quasimodular forms. This generating function is defined by universal polynomials, which can be viewed as a virtual counting of nodal curves, especially when the line bundle is not ample enough. In addition, Göttsche observed that the generating function is completely determined by the number of nodal curves on $\mathbb{P}^2$ and K3 surfaces; thus it can be computed by the Severi degrees and quasimodular forms. A reformulation of the Göttsche-Yau-Zaslow formula gives the generating function on the number of genus g curves ([**Gö**], remark 2.6) and this reformulation has been verified by Bryan and Leung [**BL**] for K3 surfaces and primitive ample line bundles.

The Severi degrees also have interesting properties. On $\mathbb{P}^2$, fixing the number of nodes and letting the degrees vary, Di Francesco and Itzykson conjectured [**FI**] that the number of plane r-nodal curves of degrees d is a polynomial in d, or the node polynomial. Recently, Fomin and Mikhalkin [**FM**] proved the polynomiality with tropical geometry and found many interesting properties of node polynomials. Block [**Bl**] generalized it to relative node polynomials and proved that there is a formal power series which specializes to all relative node polynomials. These results suggest that enumerating curves with broader conditions may possess a structure that has not been discovered.

We add that a symplectic proof to Göttsche's conjecture was given by A.K. Liu ([**Liu1**], [**Liu2**]), based on the work of Taubes on the equivalence of Seiberg-Witten theory and Gromov-Witten theory. Recently another proof was also found by Kool-Shende-Thomas [**KST**], using the BPS calculus and the computation of tautological integrals on Hilbert schemes by Ellingsrud, Göttsche and Lehn [**EGL**].

1.3. Multiplicative structure. The Göttsche-Yau-Zaslow formula expressed the generating function in terms of quasimodular forms and two unknown series $B_1(q)$ and $B_2(q)$. If the canonical divisor of a surface S is numerically trivial, only the quasimodular forms appear in the generating function. Consequently, the generating function for K3 surfaces and abelian surfaces is known. The unknown series $B_1(q)$ and $B_2(q)$ can be determined using Caporaso and Harris' recursive formulas of the Severi degrees of $\mathbb{P}^2$. Göttsche has computed the coefficients of $B_1(q)$ and $B_2(q)$ up to degree 28. However, we are still unable to find the closed forms for $B_1(q)$ and $B_2(q)$.

Instead of indexing on $(DG_2)^r$, we can simply use x^r to define another generating function

$$T(S,L) = \sum_{r\geq 0} T_r(L^2, LK, c_1(S)^2, c_2(S))x^r.$$

Because all coefficients are universal, $T(S,L)$ is a universal power series. Moreover, $T(S,L)$ is multiplicative:

THEOREM 1.3 ([**Gö**], proposition 2.3). *Assuming the numbers of nodal curves are given by universal polynomials, then there exist universal power series A_1, A_2, A_3, A_4 in $\mathbb{Q}[[x]]^\times$* [1] *such that the generating function has the form*

$$T(S,L) = A_1^{L^2} A_2^{LK_S} A_3^{c_1(S)^2} A_4^{c_2(S)}.$$

While Göttsche proved this theorem by considering disjoint union of surfaces, we will give a different proof using algebraic cobordism in Section 4. The new proof plays a central role in our approach because it demonstrates that universality is a result of algebraic cobordism structure.

The coefficients of A_i can be determined by Caporaso-Harris [**CH**] and Vakil's [**Vakil**] recursive formulas on $\mathbb{P}^2$ and $\mathbb{P}^1 \times \mathbb{P}^1$ but the closed forms are unknown. Note the generating functions in Theorems 1.2 and 1.3 can be defined for all line bundles even if they are trivial or not effective. In this setting, the r-th coefficient of $T(S,L)$ equals the number of r-nodal curves of $[S,L]$ if L is sufficiently ample relative to r. Therefore, usually only a finite number of initial coefficients honestly represent the number of nodal curves, and after that the coefficients are virtual and hard to what they mean. This difficulty will be overcome by depicting the universal polynomials as intersection numbers $d_r(S,L)$ and working with $d_r(S,L)$ directly.

[1] $\mathbb{Q}[[x]]^\times$ is the group of units in $\mathbb{Q}[[x]]$ and the group action is defined by multiplication of power series.

1.4. Approach. In this article we prove Theorem 1.3 first, and derive other theorems from this theorem. The main ingredients in our proof consist of the algebraic cobordism group $\omega_{2,1}$, the enumerative number $d_r(S, L)$ and the moduli stack of families of ideal sheaves. These techniques are developed in order to study the degeneration of line bundles on surfaces, and to derive a degeneration formula for the generating functions. As a consequence, we will prove when L is $(5r-1)$-very ample, the number of nodal curves in $|L|$ on S only depend on the class of $[S, L]$ in $\omega_{2,1}$, which can be computed by L^2, LK, $c_1(S)^2$ and $c_2(S)$.

The algebraic cobordism theory has been developed by Levine and Pandharipande [**LP**]. They call

$$[X_0] - [X_1] - [X_2] + [X_3]$$

a *double point relation* if there exists a flat family of projective schemes $\pi : \mathcal{X} \to \mathbb{P}^1$ satisfying the following properties: firstly, $\mathcal{X}$ is smooth and X_0 is the fiber over $0 \in \mathbb{P}^1$ and is smooth. Secondly, the fiber over $\infty \in \mathbb{P}^1$ is the union of two smooth components X_1 and X_2 intersecting transversally along a smooth divisor D. Thirdly, $X_3 = \mathbb{P}(1_D \oplus N_{X_1/D})$ is a $\mathbb{P}^1$ bundle over D. Define the addition of two schemes to be the disjoint union and the multiplication to be the Cartesian product. The algebraic cobordism ring is defined to be the ring generated by all smooth projective schemes modulo the double point relation.

Since the problem of counting nodal curves is about a surface S and a line bundle L on S, we generalize Levine and Pandharipande's construction to pairs of line bundles on surfaces. Let L_i be line bundles on X_i, we call

$$[X_0, L_0] - [X_1, L_1] - [X_2, L_2] + [X_3, L_3]$$

an *double point relation* if $[X_0]-[X_1]-[X_2]+[X_3]$ is a double point relation, and there exists a line bundle $\mathcal{L}$ on $\mathcal{X}$ such that $L_i = \mathcal{L}|_{X_i}$ for $i = 0, 1, 2$; $L_3 = \eta^*(\mathcal{L}|_D)$ where $\eta : X_3 \to D$ is the projection.

In [**LP**], Levine and Pandharipande defined the *algebraic cobordism group of surfaces and line bundles* $\omega_{2,1}$ to be the vector space over $\mathbb{Q}$ spanned by all pairs $[S, L]$ modulo all double point relations. The subscript $(2, 1)$ captures the dimension of surfaces and the rank of line bundles. In Section 2, we prove $\omega_{2,1}$ is a four-dimensional vector space over $\mathbb{Q}$ and $(L^2, LK, c_1(S)^2, c_2(S))$ induces the isomorphism from $\omega_{2,1}$ to $\mathbb{Q}^4$. Consequently, the class of $[S, L]$ in $\omega_{2,1}$ is linear in $L^2, LK, c_1(S)^2$ and $c_2(S)$. Two important bases of $\omega_{2,1}$ are

$$\{[\mathbb{P}^2, \mathcal{O}], [\mathbb{P}^2, \mathcal{O}(1)], [\mathbb{P}^1 \times \mathbb{P}^1, \mathcal{O}], [\mathbb{P}^1 \times \mathbb{P}^1, \mathcal{O}(1,0)]\} \quad \text{and}$$
$$\{[\mathbb{P}^2, \mathcal{O}], [\mathbb{P}^2, \mathcal{O}(1)], [S_1, L_1], [S_2, L_2]\},$$

where S_i are K3 surfaces and L_i are primitive classes on S_i with $L_1^2 \neq L_2^2$.

In general, for pairs of vector bundles of rank k on smooth projective scheme of dimension n, one can define the algebraic cobordism group $\omega_{n,k}$.

Lee and Pandharipande [**LeeP**] have studied the structure of $\omega_{n,k}$. Our result about $\omega_{2,1}$, although worked out independently, is a special case of [**LeeP**].

Another important ingredient of our proof is the enumerative number $d_r(S,L)$. This number is defined by Göttsche ([**Gö**]) as the intersection number of a closed subscheme W^{3r} of $S^{[3r]}$ and the $2r^{th}$ Chern class of the tautological bundle of L on $S^{[3r]}$, i.e.

$$d_r(S,L) = \int_{W^{3r}} c_{2r}(L^{[3r]}).$$

He proved that when L is $(5r-1)$-very ample, $d_r(S,L)$ equals the number of r-nodal curves in $[S,L]$. Although it is difficult to compute $d_r(S,L)$ directly, we believe it is the correct object to investigate for two reasons.

First, when a pair of smooth surface and ample line bundle degenerates to the singular fiber, ampleness of the line bundle is usually not preserved. Without ampleness, the universal polynomial $T_r(L^2, LK, c_1(S)^2, c_2(S))$ do not necessarily equal to the number of r-nodal curves in $[S,L]$. On the contrary, the numbers $d_r(S,L)$ can be defined for every line bundle, even if it is not ample. In the end, we will show that for every pair $[S,L]$, $d_r(S,L)$ is the universal polynomial $T_r(L^2, LK, c_1(S)^2, c_2(S))$.

Second, since $d_r(S,L)$ is defined as an intersection number, it has many good properties under degeneration. Suppose $\pi : \mathcal{X} \to \mathbb{P}^1$ defines a double point relation and U is a Zariski open set of $\mathbb{P}^1$ such that all fibers are smooth except $\pi^{-1}(\infty)$. Jun Li and Baosen Wu [**LW**] constructed a moduli stack $\mathcal{X}^{[n]} \to U$, which is the moduli stack of rank one stable relative ideal sheaves in the family $\mathcal{X}_U = \mathcal{X} \times_{\mathbb{P}^1} U \to U$. For each integer n, the moduli stack $\mathcal{X}^{[n]} \to U$ can be viewed as a family of Hilbert schemes of n points on fibers. The generic fiber of $\mathcal{X}^{[n]} \to U$ at $\infty \neq t \in U$ is the Hilbert scheme $X_t^{[n]}$ and the special fiber is the union of products of relative Hilbert schemes

$$\bigcup_{k=0}^{n} (X_1/D)^{[k]} \times (X_2/D)^{[n-k]}.$$

For each $r \in \mathbb{N}$, recall $d_r(S,L)$ is defined to be $\int_{W^{3r}} c_{2r}(L^{[3r]})$. The closed subscheme W^{3r} in $S^{[3r]}$ can be extended globally to a family of closed subschemes $\mathcal{W}^{3r}$ in $\mathcal{X}^{[3r]}$. In addition, if $\mathcal{L}$ is a line bundle on $\mathcal{X}$, then its restriction on U similarly defines a tautological bundle $\mathcal{L}^{[n]}$ on $\mathcal{X}^{[n]}$ for each n. Thus on $\mathcal{X}^{[3r]}$, the intersection of $c_{2r}(\mathcal{L}^{[3r]})$ and $\mathcal{W}^{[3r]}$ defines a family of zero cycles. Consider a double point relation $[X_0,L_0]-[X_1,L_1]-[X_2,L_2]+[X_3,L_3]$. By rational equivalence of the fibers of W^{0} over U and ∞, we show the generating function

$$\phi(S,L)(x) = \sum_{r=0}^{\infty} d_r(S,L)x^r$$

satisfies the following degeneration formula:

$$\phi(X_0, L_0) = \phi(X_1/D, L_1)\phi(X_2/D, L_2),$$

where $\phi(X_i/D, L_i)$ is the "relative" generating function.

The relation of absolute generating function ϕ can be computed by applying the degeneration formula on several families to eliminate the relative functions, which is

$$\phi(X_0, L_0) = \frac{\phi(X_1, L_1)\phi(X_2, L_2)}{\phi(X_3, L_3)}.$$

Thus ϕ induces a group homomorphism from the algebraic cobordism group $\omega_{2,1}$ to $(\mathbb{Q}[[x]]^\times, \cdot)$. Consequently, Theorem 1.3 is proved and $d_r(S, L)$ equals the universal polynomial $T_r(L^2, LK, c_1(S)^2, c_2(S))$ for all smooth projective surfaces S and line bundles L. Since $d_r(S, L)$ equals the number of r-nodal curves when L is $(5r-1)$-very ample, Theorem 1.1 is proved as a corollary.

The generating function in Theorem 1.2 is

$$\gamma(S, L)(q) = \sum_{r\in\mathbb{Z}} T_r(L^2, LK, c_1(S)^2, c_2(S))(DG_2(\tau))^r = \phi(S, L)(DG_2).$$

Therefore this generating function also induces a homomorphism from $\omega_{2,1}$ to $\mathbb{Q}[[q]]^\times$. Since Bryan and Leung [**BL**] have computed $\gamma(S, L)(q)$ on generic K3 surfaces and primitive classes, we shall use a different basis

$$\{[\mathbb{P}^2, \mathcal{O}], [\mathbb{P}^2, \mathcal{O}(1)], [S_1, L_1], [S_2, L_2]\}$$

of $\omega_{2,1}$, where S_i are K3 surfaces and L_i are primitive classes on S_i with $L_1^2 \neq L_2^2$. By Theorem 1.3, $\gamma(S, L)(q)$ is a weighted product of $\gamma(\mathbb{P}^2, \mathcal{O})$, $\gamma(\mathbb{P}^2, \mathcal{O}(1))$, $\gamma(S_1, L_1)$ and $\gamma(S_2, L_2)$. This proves Theorem 1.2. See Section 4 for more details about proofs and computation.

1.5. Outline. In Section 2 we briefly review the construction of two algebraic cobordism theories. In particular we construct the algebraic cobordism group $\omega_{2,1}$ and study its structure.

Section 3 is dedicated to the enumerative number $d_r(S, L)$ and its generating function $\phi(S, L)(x)$. The main result in this section is a degeneration formula about $\phi(S, L)(x)$ for pairs satisfying a double point relation.

Finally, in Section 4 we combine the techniques developed in Sections 2 and 3 to prove Theorems 1.1, 1.2 and 1.3. We express the infinite series A_i and B_i as the weighted product of generating functions on $\mathbb{P}^2$, $\mathbb{P}^1 \times \mathbb{P}^1$, and K3 surfaces.

1.6. Acknowledgments. I am indebted to my advisor Jun Li for his supervision, advice, and guidance. His contribution of time, ideas, and patience is invaluable. Special thanks also to Ravi Vakil for suggestions and help in all aspects. Moreover, I thank Lothar Göttsche for the explanation of his approach and Rahul Pandharipande for useful suggestions. I am also grateful to Professor Shing-Tung Yau for his support and encouragements.

1.7. Notation and Convention.

(1) All surfaces are assumed to be complex, projective, smooth and algebraic.
(2) An r-nodal curve is a reduced connected curve that has exactly r nodes and no other types of singularity.
(3) We always denote by $[S, L]$ a pair of a smooth projective surface S and a line bundle L on S.
(4) The number of r-nodal curves in a pair $[S, L]$ means the number of r-nodal curves in a generic r-dimensional linear system of $|L|$ on S.

2. Algebraic Cobordism

2.1. Introduction. In this section, we assume k be a field of characteristic 0. Let $\mathbf{Sch}_k$ denote the category of separated schemes of finite type over k and $\mathbf{Sm}_k$ denote its full subcategory consisting of smooth quasi-projective k-schemes.

The theory of algebraic cobordism Ω^* was first constructed by Levine and Morel [**LM**], which lifts the complex cobordism theory to algebraic varieties over a field of characteristic zero. They proved that Ω^* is a universal oriented cohomology theory on $\mathbf{Sm}_k$ endowed with a universal formal group law. Other examples of oriented cohomology include the Chow ring $X \to \mathbf{CH}^*(X)$, the Grothendieck group of coherent sheaves joint with β, β^{-1}: $X \to K^0(X)[\beta, \beta^{-1}]$. The universality of Ω^* implies there are induced morphisms of oriented cohomology theory

$$\Omega^* \otimes_{\mathbb{L}^*} \mathbb{Z} \to \mathbf{CH}^*$$
$$\Omega^* \otimes_{\mathbb{L}^*} \mathbb{Z}[\beta, \beta^{-1}] \to K^0[\beta, \beta^{-1}]$$

which are in fact isomorphisms. Thus $\mathbf{CH}^*$ and $K^0[\beta, \beta^{-1}]$ can be obtained by changing the formal group laws on Ω^*, which is done by tensoring factors $\mathbb{Z}$ and $\mathbb{Z}[\beta, \beta^{-1}]$.

When Levine and Morel constructed Ω^*, they first identified the homology theory Ω_* on $\mathbf{Sch}_k$ and identify two theories by $\Omega^n(X) = \bigoplus \Omega_{d_\alpha - n}(X_\alpha)$ where $X = \cup X_\alpha$ is the union of irreducible components and the dimension of (X_α) is d_α. $\Omega_*(X)$ are generated by cobordism cycles $[f : Y \to X, L_1, L_2, \ldots, L_r]$ where Y is in $\mathbf{Sm}_k$, f is projective and L_i's are line bundles over Y. Define the degree of such a cobordism cycle to be $\dim_k(Y) - r$. Then $\Omega_*(X)$ is defined as the free abelian group generated by cobordism cycles modulo some natural relations. Briefing speaking, if L is a line bundle on X, we hope to have

$$\begin{aligned} c_1(L)[f : Y \to X, L_1, L_2, \ldots, L_r] &:= [f : Y \to X, L_1, L_2, \ldots, L_r, f^*L] \\ &= [f \circ i : Z \to X, L_1, L_2, \ldots, L_r], \end{aligned}$$

where $i : Z \to Y$ is a closed subvariety of the zeroes of s and s is a section of L transversal to the zero section. Thus the relations in $\Omega_*(X)$ are imposed to make the equality always hold, and the first Chern class satisfies projective bundle formula, homotopy invariance, nilpotence, and formal group law axioms.

Instead of dealing with a large number of generators and relations, Levine and Pandharipande constructed another algebraic cobordism theory ω_* by allowing only one type of relation called the double point degeneration. More precisely, they consider projective morphisms

$$\pi : Y \to X \times \mathbb{P}^1$$

and composition

$$\pi_2 = p_2 \circ \pi : Y \to \mathbb{P}^1$$

where $X \in \mathbf{Sch}_k$ and Y is a smooth quasi-projective scheme of pure dimension. Assuming $\pi_2^{-1}(\infty) = A \cup B$, where A and B are smooth Cartier divisors intersecting transversely at $D = A \cap B$,

$$X_3 := \mathbb{P}(\mathcal{O}_D \otimes N_{A/D}) \cong \mathbb{P}(N_{B/D} \otimes \mathcal{O}_D)$$

is a $\mathbb{P}^1$ bundle over D, and Y_0 is the fiber of π_2 over 0. The double point relation over X defined by π is

$$[Y_0 \to X] - [A \to X] - [B \to X] + [X_3 \to X].$$

Let $\mathcal{M}(X)^+$ be the free additive group generated by $[M \to X]$, where M is a quasi-projective smooth scheme and the morphism is projective. Denote $\mathcal{R}(X)$ to be the subgroup generated by all double point relations over X and $\omega_*(X) = \mathcal{M}(X)^+/\mathcal{R}(X)$. Then mapping $[M \to X]$ to itself descends to a morphism between two algebraic cobordism theories:

THEOREM 2.1 ([**LP**]). *There is a canonical isomorphism*

$$\omega_*(X) \cong \Omega_*(X).$$

As a result, ω_* satisfies all properties of Ω_*. In particular, if k is a field of characteristic 0, we denote $\Omega(k)$ to be $\Omega_*(\text{Spec } k)$ and $\omega_*(k)$ to be $\omega_*(\text{Spec } k)$ in short. Levine and Morel showed that $\Omega_*(k)$ is isomorphic to the Lazard ring $\mathbb{L}_*$; consequently also isomorphic to $\omega_*(k)$. Furthermore, if k admits an embedding to $\mathbb{C}$, then

$$\begin{array}{ccccc} \omega_*(k) \to & & \Omega_*(k) \to & & MU_{2*} \cong \mathbb{L}_* \\ [X \to \text{Spec}k] & \to & [X \to \text{Spec}k] & \to & [X \times_{\text{Spec}k} \text{Spec}\mathbb{C}] \end{array}$$

are isomorphisms, where the last morphism is Quillen's isomorphism. Since $MU_{2*} \otimes_{\mathbb{Z}} \mathbb{Q} \cong \mathbb{L}_* \otimes_{\mathbb{Z}} \mathbb{Q}$ has a basis formed by products of projective spaces, it follows that

COROLLARY 2.2.

$$\Omega_*(k) \otimes_{\mathbb{Z}} \mathbb{Q} \cong \omega_*(k) \otimes_{\mathbb{Z}} \mathbb{Q} \cong \bigoplus_{\lambda=(\lambda_1,\ldots,\lambda_r)} \mathbb{Q}[\mathbb{P}^{\lambda_1} \times \cdots \times \mathbb{P}^{\lambda_r}]$$

where the index λ belongs to $\mathbb{N}^r$ for some positive integer r. Moreover, the class of a smooth scheme X is uniquely determined by all Chern numbers of X, in the same way as complex cobordism theory

In particular, $\omega_2(k)$ is generated by $\mathbb{P}^2$ and $\mathbb{P}^1 \times \mathbb{P}^1$ over $\mathbb{Q}$, i.e. all smooth projective surfaces, up to a multiple, can be degenerated to the sum of $\mathbb{P}^2$ and $\mathbb{P}^1 \times \mathbb{P}^1$ using double point relations. In addition, the coefficients of the sum are uniquely determined by the Chern numbers of the surface.

For our purpose of counting nodal curves, degenerating surfaces alone are not sufficient. Instead, we consider the degeneration of pairs $[S, L]$ where S is a smooth projective surface and L is a line bundle on S.

DEFINITION 2.1. Suppose $[X_i, L_i]$ are pairs of smooth projective surfaces over k and line bundles for $i = 0, 1, 2, 3$. The double point relation for such pairs is defined by

$$[X_0, L_0] = [X_1, L_1] + [X_2, L_2] - [X_3, L_3] \tag{2.1}$$

with the assumption that there exists a flat family of surfaces $\pi : \mathcal{X} \to \mathbb{P}^1$ and a line bundle $\mathcal{L}$ on $\mathcal{X}$ which satisfies the following properties:

(1) the family $\pi : \mathcal{X} \to \mathbb{P}^1$ gives a double point relation on surfaces;
(2) L_i is the restriction of $\mathcal{L}$ on X_i for $i = 0, 1, 2$;
(3) L_3 is the pullback of $\mathcal{L}|_D$ via the projection $X_3 \to D$.

REMARK. We use X_1/D and X_2/D to emphasize the divisor D when discussing the relative geometry on X_i.

Let $M(k)$ be the $\mathbb{Q}$-vector space[2] spanned by pairs of smooth projective surfaces and line bundles, and let $R(k)$ be the subgroup spanned by all double point relations for pairs. We define the *algebraic cobordism group of surfaces and line bundles* to be

$$\omega_{2,1}(k) = M(k)/R(k).$$

Because the double point relation on surfaces can be viewed as a double point relation on pairs of surfaces with trivial line bundle, $\omega_2(k)$ naturally embeds into $\omega_{2,1}(k)$

By definition, $\omega_{2,1}(k)$ is a vector space over $\mathbb{Q}$. We are interested in finding its dimension, bases, and invariants of this degeneration theory.

PROPOSITION 2.3. *Let S be a smooth projective surface over k and L be a line bundle on S, then*

$$[S, L] = a_1[\mathbb{P}^2, \mathcal{O}] + a_2[\mathbb{P}^2, \mathcal{O}(1)] + a_3[\mathbb{P}^1 \times \mathbb{P}^1, \mathcal{O}] + a_4[\mathbb{P}^1 \times \mathbb{P}^1, \mathcal{O}(1,0)] \tag{2.2}$$

---

[2] We use $\mathbb{Q}$ here because it is enough for our purpose and for simplicity. In general, the coefficients can be $\mathbb{Z}$ as discussed in [**LP**].

in $\omega_{2,1}(k)$, *where*

$$a_1 = -L^2 + \frac{c_1(S)^2 + c_2(S)}{3} - c_2(S), \qquad a_2 = L^2,$$
$$a_3 = L^2 + \frac{LK + L^2}{2} - \frac{c_1(S)^2 + c_2(S)}{4} + c_2(S), \quad a_4 = -L^2 - \frac{LK + L^2}{2}.$$

In other words, $\omega_{2,1}(k)$ *is a four dimensional vector space spanned by four elements* $[\mathbb{P}^2, \mathcal{O}]$, $[\mathbb{P}^2, \mathcal{O}(1)]$, $[\mathbb{P}^1 \times \mathbb{P}^1, \mathcal{O}]$ *and* $[\mathbb{P}^1 \times \mathbb{P}^1, \mathcal{O}(1,0)]$. *Moreover,* $(L^2, LK, c_1(S)^2, c_2(S))$ *defines an isomorphism from* $\omega_{2,1}(k)$ *to* $\mathbb{Q}^4$.

Remark. Proposition 2.3 implies the class of $[S, L]$ in $\omega_{2,1}$ is uniquely determined by degree two Chern polynomials of $c_1(S)$, $c_2(S)$ and $c_1(L)$ which is a generalization of the case of schemes. Noether's formula states

$$\chi(\mathcal{O}_S) = \frac{1}{12}(c_1(S)^2 + c_2(S))$$

and thus $c_1(S)^2 + c_2(S)$ is divisible by three and four. In addition, the Riemann-Roch formula for L is

$$\chi(L) = \chi(\mathcal{O}_S) + \frac{1}{2}(L^2 - LK),$$

and therefore $L^2 - LK$ is divisible by two. Therefore all coefficients in (2.2) are integers.

In [**Tz**], we give a elementary proof of Proposition 2.3, which starts with the following observation:

Lemma 2.4. *Let* C *be a smooth curve in* S, L *be a line bundle on* S, N *be the normal bundle of* C *in* S *and* $\eta : \mathbb{P}_C(N \oplus \mathcal{O}_C) \to C$ *be the structure map. Then*

$$[S, L] = [S, L \otimes \mathcal{O}_S(-C)] + \textit{pairs on } \mathbb{P}_C(N \oplus \mathcal{O}_C)$$

in $\omega_{2,1}(k)$.

Consequently, we are allowed to add effective divisors freely to the line bundle. The remaining terms are on ruled surfaces which have been classified. As a result Proposition 2.3 can be proved in the following steps.

(1) Write L as the $\mathcal{O}(D_1 - D_2)$, where $\mathcal{O}(D_i)$ are very ample and D_i are smooth divisors. Then our observation implies $[S, L]$ is the sum of $[S, \mathcal{O}]$ plus pairs on ruled surfaces in the form $\mathbb{P}_C(N \oplus \mathcal{O}_C)$.
(2) Degeneration of $[S, \mathcal{O}]$ is the same as degenerating $[S]$ in $\omega_2(k)$, thus $[S, \mathcal{O}]$ can be written as the sum of several $[\mathbb{P}^2, \mathcal{O}]$ and $[\mathbb{P}^1 \times \mathbb{P}^1, \mathcal{O}]$.
(3) Pairs on $\mathbb{P}_C(N \ominus \mathcal{O}_C)$ can be further simplified by modifying the line bundle N and degenerating the base curve C. Then we write explicit double point relations to conclude these pairs are generated by $[\mathbb{P}^2, \mathcal{O}(1)]$, $[\mathbb{P}^1 \times \mathbb{P}^1, \mathcal{O}]$ and $[\mathbb{P}^1 \times \mathbb{P}^1, \mathcal{O}(1,0)]$.

(4) Finally, we show L^2, LK, $c_1(S)^2$ and $c_2(S)$ are preserved in double point relations, thus are invariants of $\omega_{2,1}(k)$. It implies $\omega_{2,1}(k)$ has four generators and four independent invariants, so the generators is actually a basis of $\omega_{2,1}(k)$.
(5) Since L^2, LK, $c_1(S)^2$ and $c_2(S)$ are invariants of $\omega_{2,1}(k)$, it can be used to compute the coefficients of equation 2.2.

REMARK.

(1) From the proof of Proposition 2.3, we know a set of four elements

$$\{[S_i, L_i] \mid i = 1, \ldots, 4\}$$

is a basis of $\omega_{2,1}$ if and only if the four vectors

$$\{(L_i^2, L_i K_{S_i}, c_1(S_i)^2, c_2(S_i)) \mid i = 1, \ldots, 4\}$$

are linearly independent over $\mathbb{Q}$.
(2) The sets $(L^2, LK, c_1(S)^2, c_2(S))$ and $(LK, \chi(L), \chi(\mathcal{O}), K^2)$ determine each other, thus $(LK, \chi(L), \chi(\mathcal{O}), K^2)$ is also a set of invariants of $\omega_{2,1}$. As a result, if for a set of four elements in $\omega_{2,1}$, the corresponding vectors $(LK, \chi(L), \chi(\mathcal{O}), K^2)$ are linearly independent, then the set is a basis of $\omega_{2,1}$.
(3) If S_1, S_2 are two K3 surfaces and L_i are primitive classes on S_i respectively, the four numbers $(LK, \chi(L), \chi(\mathcal{O}), K^2)$ of

$$\mathfrak{B} = \{[\mathbb{P}^2, \mathcal{O}], [\mathbb{P}^2, \mathcal{O}(1)], [S_1, L_1], [S_2, L_2]\} \quad \text{are}$$

$$(0,1,1,9), (-3,3,1,9), \left(0, 2 + \frac{L_1^2}{2}, 2, 0\right), \left(0, 2 + \frac{L_2^2}{2}, 2, 0\right).$$

It follows that $\mathfrak{B}$ is a basis if and only if $L_1^2 \neq L_2^2$.

In general, the algebraic cobordism theory $\omega_{n,r}$ for smooth schemes of dimension n and vector bundles of rank r have been constructed analogously by Y.P. Lee and Pandharipande [**LeeP**] for all nonnegative n, r. They proved that the invariants of $\omega_{n,r}$ are all possible degree n monomials of the Chern classes of the scheme and the vector bundle. Furthermore, there exists a basis consisting of products of projective spaces pairs with direct sum of selected pullback of $\mathcal{O}(1)$'s.

3. Degeneration Formula

3.1. The enumerative number $d_r(S, L)$. In this section we use the enumerative number $d_r(S, L)$ to study the number of nodal curves. This number $d_r(S, L)$ was introduced by Göttsche [**Gö**] for pairs of smooth projective surface and line bundle $[S, L]$, and he proved that $d_r(S, L)$ equals the number of r-nodal curves on $[S, L]$ if L is $(5r-1)$-very ample. The goal of this section is to derive a degeneration formula for the generating function

$$\phi(S,L)(x) = \sum_{r=0}^{\infty} d_r(S,L)x^r$$

for pairs satisfying double point relations.

Let $S^{[n]}$ be the Hilbert scheme of n points on S, and let $Z_n \subset S \times S^{[n]}$ be the universal closed subscheme with projections

$$p_n : Z_n \to S,\ q_n : Z_n \to S^{[n]}.$$

Define $L^{[n]} = (q_n)_*(p_n)^*L$. Because q_n is finite and flat, $L^{[n]}$ is a vector bundle of rank n on $S^{[n]}$. Göttsche [**Gö**] suggested the following approach, which interprets the number of nodal curves as intersection numbers on Hilbert schemes:

Definition 3.1 ([**Gö**], definition 5.1). Let W_0^{3r} be the locally closed subset

$$\left\{ \coprod_{i=1}^{r} \operatorname{Spec}(\mathcal{O}_{S,x_i}/m_{S,x_i}^2) \,\middle|\, x_i \text{ are distinct closed points on } S \right\}$$

and $W^{3r} \subset S^{[3r]}$ be the closure of W_0^{3r} (with the reduced induced structure). It is easy to see that W^{3r} is birational to $S^{[r]}$. Define

$$d_r(S,L) = \int_{W^{3r}} c_{2r}(L^{[3r]}).$$

For simplicity, define $d_0(S,L) = 1$ because the number of 0-nodal curves, which by definition are smooth, in a zero-dimensional linear system is one.

Definition 3.2. We call a line bundle L *k-very ample* if for every zero-dimensional subscheme $\xi \subset S$ of length $k+1$, the natural map $H^0(S,L) \to H^0(\xi, L \otimes \mathcal{O}_\xi)$ is surjective.

If L and M are very ample, then $L^{\otimes k} \otimes M^{\otimes l}$ is $(k+l)$-very ample. In particular, very ampleness implies 1-very ampleness.

We quote a result of Göttsche below:

Proposition 3.1 ([**Gö**], proposition 5.2). *Assume S is a smooth algebraic surface and L is a $(5r-1)$-very ample line bundle on S, then a general r-dimensional sublinear system $V \subset |L|$ contains $d_r(S,L)$ curves with precisely r-nodes as singularities.*

3.2. Degeneration Formula. In Section 1.4 we explained why $d_r(S,L)$ behaves better than the actual number of nodal curves in flat families. The goal of this section is to derive a degeneration formula for $d_r(S,L)$. More precisely, we will show that if

$$[X_0, L_0] = [X_1, L_1] + [X_2, L_2] - [X_3, L_3]$$

is a double point relation, then then number of nodal curves on $[X_0, L_0]$ can be determined by the numbers on $[X_1, L_1]$, $[X_2, L_2]$ and $[X_3, L_3]$. Thus it is necessary to treat the number of curves with varied number of nodes together by considering the generating function

$$\phi(S, L)(x) = \sum_{r=0}^{\infty} d_r(S, L)x^r.$$

The following theorem is the main result of this section:

PROPOSITION 3.2. *Suppose* $[X_0, L_0] = [X_1, L_1] + [X_2, L_2] - [X_3, L_3]$ *is a double point relation. Then*

$$\phi(X_0, L_0) = \frac{\phi(X_1, L_1) \cdot \phi(X_2, L_2)}{\phi(X_3, L_3)}.$$

In other words, ϕ *is a homomorphism from* $\omega_{2,1}$ *to* $(\mathbb{Q}[[x]]^{\times}, \ \cdot \ )$.

3.3. Moduli stack of relative ideal sheaves. The key tool is Jun Li and Baosen Wu's [**LW**] construction of the moduli stack of stable relative ideal sheaves. Similar construction has been applied successfully to prove the degeneration formulas of Gromov-Witten invariants [**Li**] and of Donalsdon-Thomas invariants [**LW**]. Our case is even simpler because we only consider families of surfaces and ideal sheaves on points on surfaces.

Let $\infty \in C$ be a specialized point and $\pi : X \to C$ be a flat projective family of schemes that satisfies

(1) X is smooth and π is smooth away from the fiber $\pi^{-1}(\infty)$;
(2) $\pi^{-1}(\infty) =: X_1 \cup_D X_2$ is a union of two irreducible smooth components X_1 and X_2 which intersect transversally along a smooth divisor D.

In [**LW**], Li and Wu defined the notion of a family of stable perfect ideal sheaves over C and constructed $\mathfrak{I}^{\Gamma}_{X/C}$, the moduli space of stable perfect ideal sheaves of type Γ of $X \to C$. To make $\mathfrak{I}^{\Gamma}_{X/C}$ a stack, one has to replace X by new spaces $X[n]$ so that X and $X[n]$ have the same smooth fiber X_t when $t \neq \infty$. Over ∞, the fiber of $X[n]$ is a semistable model

$$X[n]_0 = X_1 \cup \Delta_1 \cup \Delta_2 \cup \cdots \Delta_{n-1} \cup X_2,$$

where $\Delta_i \cong \mathbb{P}_D(\mathcal{O}_D \oplus N_{X_1/D})$. The objects of $\mathfrak{I}^{\Gamma}_{X/C}$ are expanded relative pairs $(\mathcal{X}/S, \mathcal{I})$ which consists of a family $\mathcal{X}$ over a C-scheme S and a family of stable ideal sheaves $\mathcal{I}$ of type Γ on $\mathcal{X}/S$. The fibers $\mathcal{X}_s$ of the family $\mathcal{X}/S$ are required to be either smooth a fiber of X/C or a semistable model $X[n]_0$ for some n. Under these settings, Li and Wu proved that the moduli space has many good properties:

THEOREM 3.3 ([**LW**]). *The moduli stack* $\mathfrak{I}^{\Gamma}_{X/C}$ *is a separated and proper Deligne-Mumford stack of finite type over* C.

In our case, $\pi : X \to C$ is a family of surfaces and $\mathcal{I}$ is a family of ideal sheaves of zero-dimensional closed subschemes of length n. For $s \in S$, when $\mathcal{X}_s$ is a smooth fiber of X/C, $\mathcal{I}_s$ is automatically perfect and stable. If $\mathcal{X}_s$ is a semistable model $X[m]_0$, then the support of $\mathcal{I}_s$ can not lie on the singular loci of $X[m]_0$. Furthermore, every component Δ_i contains at least one point of the zero-set of $\mathcal{I}_s$ (otherwise this component can be contracted).

The resulted moduli space $\mathfrak{I}^n_{X/C}$ a separated and proper Deligne-Mumford stack of finite type over C, and the fibers of the projection $\pi^{[n]} : \mathfrak{I}^n_{X/C} \to C$ can be described as follows:

(1) The fiber over ∞ is the union of products

$$\cup_{k=0}^n (X_1/D)^{[k]} \times (X_2/D)^{[n-k]}$$

for all possible $n \geq k \geq 0$. $(X_i/D)^{[n_i]}$ are the moduli spaces of stable relative ideal sheaves of n_i points on X_i/D. $(X_i/D)^{[0]} = pt$. They are also separated and proper Deligne-Mumford stacks (see [**LW**]). We denote this fiber by $(X_1 \cup_D X_2)^{[n]}$.

(2) $\mathfrak{I}^n_{X/C}$ is smooth and $\pi^{[n]}$ is smooth away from the fiber over ∞.

(3) When $t \neq \infty$, the smooth fiber of $\mathfrak{I}^n_{X/C}$ over t equals $X_t^{[n]}$, the Hilbert schemes of n points on X_t.

Therefore, $\mathfrak{I}^n_{X/C}$ can also be viewed as a family of Hilbert schemes of n points on X/C.

Suppose a family of surfaces $\mathcal{X} \to \mathbb{P}^1$ and line bundle $\mathcal{L}$ on $\mathcal{X}$ gives a double point relation of line bundles on surfaces. This family $\mathcal{X}$ may have multiple singular fibers and only the fiber of ∞ is required to be two components intersection transversally. Therefore Li and Wu's construction can be applied only on an open subset of $\mathbb{P}^1$.

Let U be a Zariski open set of $\mathbb{P}^1$ obtained by deleting those points with singular fibers except ∞, i.e. set-theoretically,

$$U = \{t \in \mathbb{P}^1 \mid \text{ the fiber } \mathcal{X}|_t \text{ is smooth } \} \cup \{\infty\}.$$

Then $\pi_U : \mathcal{X}_U := \mathcal{X} \times_{\mathbb{P}^1} U \to U$ is a family of surfaces with only one singular fiber X_∞.

Write $\mathfrak{I}^n_{\mathcal{X}_U/U}$ as $\mathcal{X}^{[n]}$. There are also universal closed subschemes over $\mathcal{X}^{[n]}$, $(X_1/D)^{[n]}$ and $(X_2/D)^{[n]}$ respectively. Similarly, one can define rank n vector bundles $\mathcal{L}^{[n]}$, $L_1^{[n]}$ and $L_2^{[n]}$ on $(X_1/D)^{[n]}$ and $(X_2/D)^{[n]}$ respectively, and closed substacks $\mathcal{W}^{3r}$, $W^{3r}_{X_0}$ and $W^{3r}_{X_i/D}$ for $i = 1, 2$ in $\mathcal{X}^{[3r]}$, $X_0^{[3r]}$ and $(X_i/D)^{[3r]}$. Let

$$d_r(X_i/D, L_i) = \int_{W^{3r}_{X_i/D}} c_{2r}(L_i^{[3r]}), \; d_0(X_i/D, L_i) = 1$$

and

$$\phi(X_i/D, L_i) = \sum_{r\geq 0} d_r(X_i/D, L_i)x^r.$$

The 1-cycle $\int_{\mathcal{W}^{3r}} c_{2r}(\mathcal{L}^{[3r]})$ restricts to fibers of $\mathcal{X}^{[n]}$ over 0 and ∞ gives two rational equivalent 0-cycles with the same degree. Therefore

PROPOSITION 3.4. *Suppose* $[X_0, L_1] - [X_1, L_1] - [X_2, L_2] + [X_3, L_3]$ *is a double point relation, then*

$$\phi(X_0, L_0) = \phi(X_1/D, L_1) \cdot \phi(X_2/D, L_2).$$

To obtain a formula for absolute generating function, the relative surfaces can be closed up by adding projective bundles.

COROLLARY 3.5. *Let* $[X, L]$ *be a pair in* $\omega_{2,1}$. *Suppose* D *is a smooth curve in* X, N *is the normal bundle of* D *in* X *and* $L \cdot D$ *is the intersection number of* $c_1(L)$ *and* D. *Recall* $\mathbb{P}_N := \mathbb{P}_D(\mathcal{O}_D \oplus N)$, *then*

$$\phi(X, L) = \phi(X/D, L) \cdot \phi(\mathbb{P}_N/D, (L \cdot D)f).$$

Now we are ready to prove Proposition 3.2

PROOF OF PROPOSITION 3.2. Apply Corollary 3.5 to $X = X_1$, X_2 and $\mathbb{P}_N$. It is easy to check the embeddings of D match perfectly and $(L_i \cdot D)f \cong (L_2 \cdot D)f \cong L_3$ on $\mathbb{P}_N$. Thus we obtain

$$\begin{aligned}
\phi(X_0, L_0) &= \phi(X_1/D, L_1) \cdot \phi(X_2/D, L_2), \\
\phi(X_1/D, L_1) \cdot \phi(\mathbb{P}_N/D_0, L_3) &= \phi(X_1, L_1), \\
\phi(X_2/D, L_2) \cdot \phi(\mathbb{P}_N/D_\infty, L_3) &= \phi(X_2, L_2), \\
\phi(\mathbb{P}_N, L_3) &= \phi(\mathbb{P}_N/D_0, L_3) \cdot \phi(\mathbb{P}_N/D_\infty, L_3).
\end{aligned}$$

Then the theorem is proved by multiplying all equations. □

4. Universality Theorems and Generating functions

4.1. Outline. In this Section, we will prove Theorems 1.1, 1.2 and 1.3 by combining the degeneration formula (Proposition 3.2) with the structure of algebraic cobordism group $\omega_{2,1}$ (Proposition 2.3).

Recall that for any smooth projective surface S and line bundle L on S, we defined and studied the enumerative number $d_r(S, L)$ and generating function

$$\phi(S, L)(x) = \sum_{r=0}^{\infty} d_r(S, L)\, x^r$$

in Section 3. By Proposition 3.2, this function ϕ induces a homomorphism from $\omega_{2,1}$ to $(\mathbb{Q}[[x]]^{\times}, \cdot)$. On the other hand, Proposition 2.3 proves that $\omega_{2,1}$ is four-dimensional and the only invariants are L^2, LK, $c_1(S)^2$ and $c_2(S)$. Combining these two results, we show that $\phi(S, L)$ only depends on these four topological numbers and has a multiplicative structure.

4.2. Proof of Theorems 1.3 and 1.1.

PROPOSITION 4.1. *There exist four series A_1, A_2, A_3 and A_4 in $\mathbb{Q}[[x]]^\times$ such that*

$$\phi(S,L)(x) = A_1^{L^2} A_2^{LK_S} A_3^{c_1(S)^2} A_4^{c_2(S)}.$$

More explicitly,

$$A_1(x) = \phi(\mathbb{P}^2,\mathcal{O})^{-1}\phi(\mathbb{P}^2,\mathcal{O}(1))\phi(\mathbb{P}^1\times\mathbb{P}^1,\mathcal{O})^{\frac{3}{2}}\phi(\mathbb{P}^1\times\mathbb{P}^1,\mathcal{O}(1,0))^{-\frac{3}{2}},$$
$$A_2(x) = \phi(\mathbb{P}^1\times\mathbb{P}^1,\mathcal{O})^{\frac{1}{2}}\phi(\mathbb{P}^1\times\mathbb{P}^1,\mathcal{O}(1,0))^{-\frac{1}{2}},$$
$$A_3(x) = \phi(\mathbb{P}^2,\mathcal{O})^{-\frac{1}{3}}\phi(\mathbb{P}^1\times\mathbb{P}^1,\mathcal{O})^{-\frac{1}{4}},$$
$$A_4(x) = \phi(\mathbb{P}^2,\mathcal{O})^{-\frac{2}{3}}\phi(\mathbb{P}^1\times\mathbb{P}^1,\mathcal{O})^{\frac{3}{4}}.$$

PROOF. By Proposition 2.3, the class of $[S,L]$ in $\omega_{2,1}$ is

(4.1)
$$[S,L] = a_1[\mathbb{P}^2,\mathcal{O}] + a_2[\mathbb{P}^2,\mathcal{O}(1)] + a_3[\mathbb{P}^1\times\mathbb{P}^1,\mathcal{O}] + a_4[\mathbb{P}^1\times\mathbb{P}^1,\mathcal{O}(1,0)]$$

where

$$a_1 = -L^2 + \frac{c_1(S)^2 + c_2(S)}{3} - c_2(S), \qquad a_2 = L^2,$$
$$a_3 = L^2 + \frac{LK+L^2}{2} - \frac{c_1(S)^2+c_2(S)}{4} + c_2(S), \quad a_4 = -L^2 - \frac{LK+L^2}{2}.$$

Since the generating function ϕ is a homomorphism from $\omega_{2,1}$ to $(\mathbb{Q}[[x]]^\times, \cdot\,)$ (Proposition 3.2), we obtain

$$\phi(S,L)(x) = \phi(\mathbb{P}^2,\mathcal{O})^{a_1}\phi(\mathbb{P}^2,\mathcal{O}(1))^{a_2}\phi(\mathbb{P}^1\times\mathbb{P}^1,\mathcal{O})^{a_3}\phi(\mathbb{P}^1\times\mathbb{P}^1,\mathcal{O}(1,0))^{a_4}.$$

Plugging in the values of a_i and grouping the functions according to L^2, LK, $c_1(S)^2$ and $c_2(S)$ complete the proof. □

Another way to prove Proposition 4.1 is using the following diagram:

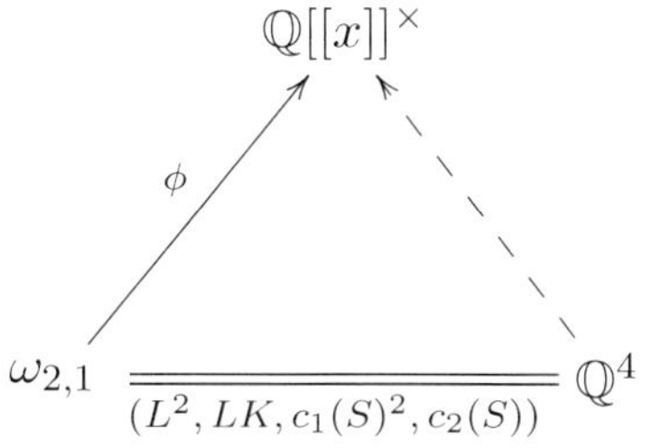

Since ϕ is a homomorphism and $\omega_{2,1} \to \mathbb{Q}^4$ is an isomorphism, they induce a homomorphism from $\mathbb{Q}^4$ to $\mathbb{Q}[[x]]^\times$. This implies $\phi(S,L)(x)$ is a universal power series in L^2, LK, $c_1(S)^2$ and $c_2(S)$. Furthermore, let the image of the standard basis $\mathbf{e_i}$ of $\mathbb{Q}^4$ be A_i; then the commutative diagram implies

$$\phi(S,L)(x) = A_1^{L^2} A_2^{LK_S} A_3^{c_1(S)^2} A_4^{c_2(S)}.$$

REMARK. One can see the same proof is still valid if we replace

$$\{[\mathbb{P}^2,\mathcal{O}],[\mathbb{P}^2,\mathcal{O}(1)],[\mathbb{P}^1\times\mathbb{P}^1,\mathcal{O}],[\mathbb{P}^1\times\mathbb{P}^1,\mathcal{O}(1,0)]\}$$

by any basis of $\omega_{2,1}$. Thus the four series $A_i's$ can also be expressed in the generating function ϕ of any basis.

COROLLARY 4.2. *The enumerative number $d_r(S,L)$ is a universal polynomial of degree r in L^2, LK, $c_1(S)^2$ and $c_2(S)$ for all projective smooth surfaces S and line bundles L on S.*

PROOF. For all pairs $[S,L]$, $d_r(S,L)$ is the coefficient of x^r in

$$\phi(S,L)(x)=A_1^{L^2}A_2^{LK_S}A_3^{c_1^2(S)}A_4^{c_2(S)}.$$

After expanding the series, one can see that the coefficient of x^r in $\phi(S,L)(x)$ is a universal polynomial of $(L^2,LK,c_1(S)^2,c_2(S))$ of degree r. □

Next, we prove Göttsche's conjecture:

THEOREM 1.1. *For every integer $r\geq 0$, there exists a universal polynomial $T_r(x,y,z,t)$ of degree r with the following property: given a smooth projective surface S and a $(5r-1)$-very ample (5-very ample if $r=1$) line bundle L on S, a general r-dimensional sublinear system of $|L|$ contains exactly $T_r(L^2,LK,c_1(S)^2,c_2(S))$ r-nodal curves.*

PROOF. $d_r(S,L)$ is always a universal polynomial of degree r for all line bundles L on S. If L is $(5r-1)$-very ample, Proposition 3.1 shows $d_r(S,L)$ equals the number of r-nodal curves in $[S,L]$. Thus the number of r-nodal curves in $[S,L]$ is given by a universal polynomial of degree r when L is $(5r-1)$-very ample. □

COROLLARY 4.3. *The generating functions $\phi(S,L)(x)$ and $T(S,L)(x)$ (defined in Section 1.3 as $\sum_{r=0}^{\infty}T_r(L^2,LK,c_1(S)^2,c_2(S))x^r$) are equal.*

THEOREM 1.3. *There exist universal power series A_1, A_2, A_3, A_4 in $\mathbb{Q}[[x]]^\times$ such that the generating function $T(S,L)$ has the form*

$$T(S,L)=A_1^{L^2}A_2^{LK_S}A_3^{c_1(S)^2}A_4^{c_2(S)}.$$

PROOF. This follows from Proposition 4.1 and Corollary 4.3. □

4.3. Proof of the Göttsche-Yau-Zaslow Formula. Let τ be on the complex upper half plane, $q=e^{2\pi i\tau}$. The Eisenstein series

$$G_2(\tau):=\frac{-1}{24}+\sum_{n>0}\left(\sum_{d|n}d\right)q^n$$

is a quasimodular form. Denote by D the differential operator $D:=\frac{1}{2\pi i}\frac{d}{d\tau}=q\frac{d}{dq}$. Since the ring of quasimodular forms is closed under differentiation,

DG_2 and D^2G_2 are also quasimodular. In addition, let

$$\Delta(\tau) = q \prod_{k>0} (1 - q^k)^{24} = \eta(\tau)^{24}$$

where $\eta(\tau)$ is the Dedekind η function, and $\Delta(\tau)$ is a modular form. We define the generating function

$$\gamma(S, L)(q) = \sum_{r=0}^{\infty} T_r(L^2, LK, c_1(S)^2, c_2(S))(DG_2(\tau))^r.$$

An equivalent form of Bryan and Leung's result ([**BL**], theorem 1.1) states that for a generic K3 surface S and a primitive class L,

$$\gamma(S, L)(q) = \frac{(DG_2(\tau)/q)^{\chi(L)}}{\Delta(\tau) D^2G_2(\tau)/q^2}.$$

More generally, the generating function $\gamma(S, L)(q)$ for all algebraic surfaces is given by

THEOREM 1.2 (Göttsche-Yau-Zaslow Formula). *There exist universal power series B_1, B_2 in q such that*

$$\sum_{r=0}^{\infty} T_r(L^2, LK, c_1(S)^2, c_2(S))(DG_2(\tau))^r$$
$$= \frac{(DG_2(\tau)/q)^{\chi(L)} B_1(q)^{K_S^2} B_2(q)^{LK_S}}{(\Delta(\tau) D^2G_2(\tau)/q^2)^{\chi(\mathcal{O}_S)/2}}.$$

PROOF. Since $(K_S^2, LK_S, \chi(L), \chi(\mathcal{O}_S))$ and $(L^2, LK, c_1(S)^2, c_2(S))$ determine each other linearly, $\phi(S, L)$ is multiplicative in $(K_S^2, LK_S, \chi(L), \chi(\mathcal{O}_S))$ and so as $\gamma(S, L)(q) = \phi(S, L)(DG_2)$. Therefore we can let

$$\gamma(S, L)(q) = B_1(q)^{K_S^2} B_2(q)^{LK_S} B_3(q)^{\chi(L)} B_4(q)^{\chi(\mathcal{O}_S)}.$$

For all generic K3 surfaces S and primitive classes L,

$$\gamma(S, L)(q) = \frac{(DG_2(\tau)/q)^{\chi(L)}}{\Delta(\tau) D^2G_2(\tau)/q^2}.$$

These pairs can achieve infinitely many distinct $\chi(L)$, and thus

$$B_3(q) = DG_2(\tau)/q \quad \text{and } B_4(q) = (\Delta(\tau) D^2G_2(\tau)/q^2)^{1/2}$$

and the result follows. □

REMARK. Although in [**BL**] they found the formula for generating function of primitive classes on K3 surfaces with Picard number one, it is not trivial to see why the coefficients equals our universal polynomials. For each r, we have to show that there exist infinitely many K3 surfaces with Picard number one and primitive classes L such that L is $(5r - 1)$-very ample and also have distinct intersection numbers L^2. It can be proved by using Lemma 5.3 in [**KP2**], which shows that on a K3 surface with Picard number one, a primitive class L is $(5r-1)$-very ample if $L^2 > 20r$. Therefore Bryan and Leung's generating function is exactly $\phi(S,L)(x)$.

There is a reformulation of Theorem 1.3 given by Göttsche ([**Gö**], remark 2.6), which is the original version of Bryan and Leung's formula.

Corollary 4.4. *For all l, m, r, define*

$$n_r^S(l,m) = T^S_{l+\chi(\mathcal{O}_S)-1-r}(2l+m,m).$$

Then

$$\sum_{l\in\mathbb{Z}} n_r^S(l,m)q^l = B_1(q)^{K_S^2} B_2(q)^m (DG_2(\tau))^r \frac{D^2G_2(\tau)}{(\Delta(\tau)D^2G_2(\tau)/q^2)^{\chi(\mathcal{O}_S)/2}}.$$

If L is sufficiently ample with respect to $\delta = \chi(L)-1-r$, then $n_r^S((L^2 - LK_S)/2, LK_S)$ counts the δ-nodal curves in a general r-codimensional sublinear system of $|L|$. This reformulation provides another (possibly better) way to find the closed form of $B_1(q)$ and $B_2(q)$.

References

[BL] J. Bryan and C. Leung, *The enumerative geometry of K3 surfaces and modular forms,* J. Amer. Math. Soc. **13** (2000), 371–410.

[Bl] F. Block, *Relative Node polynomials for plane curves,* preprint, arXiv:1009.5063.

[Ch] Y. Choi, *Severi Degrees in Cogenus 4,* preprint, arXiv:alg-geom/9601013.

[CH] L. Caporaso and J. Harris, *Counting plane curves of any genus,* Invent. Math. **131** No. 2 (1998), 345–292.

[EGL] G.Ellingsrud, L. Göttsche and M. Lehn, *On the cobordism class of the Hilbert scheme of a surface,* Jour. Alg. Geom. **10** (2000), 81–100.

[FM] S. Fomin and G. Mikhalkin, *Labeled floor diagrams for plane curves,* J. Europ. Math. Soc. **12** (2010), 1453–1496.

[FI] P. Di Francesco and C. Itzykson, *Quantum intersection rings* in The Moduli Space of Curves, Birkhäuser, (1995), 81–148.

[Fu] W. Fulton, *Intersection Theory,* A series of Modern Surveys in Mathematics, Springer-Verlag, (1984).

[Gö] L. Göttsche, *A Conjectural Generating Function for Numbers of Curves on Surfaces,* Communications in Mathematical Physics. **196** (1998), 523–533.

[Ha] R. Hartshorne, *Algebraic Geometry,* Graduate Text in Mathematics, Springer-Verlag 52.

[HP] J. Harris and R. Pandharipande, *Severi degrees in Cogenus 3,* preprint, arXiv:alg-geom/9504003.

[KM] M. Kontsevich and Y. Manin, *Gromov-Witten classes, quantum cohomology, and enumerative geometry,* Comm. Math. Phys. **164** (1994), 525–562.

[KP1] S. Kleiman and R. Piene, *Enumerating singular curves on surfaces,* Cont. Math. **241** (1999), 209–238.

[KP2] S. Kleiman and R. Piene, *Node polynomials for families: methods and applications,* Math. Nachr. **271** (2004), 69–90.

[KST] M. Kool, V. Shende and R. Thomas, *A short proof of the Göttsche conjecture,* Geom. Topol. **15**, No. 1 (2011), 397–406.

[KZ] M. Kaneko and D. Zagier, *A generalized Jacobi theta function and quasimodular forms* in The moduli space of curves, eds. R. Dijkgraaf, C. Faber, G. van der Geer, Boston: Birkhäuser, (1995), 165–172.

[Li] J. Li, *Relative Gromov-Witten invariants and a degeneration formula of Gromov-Witten invariants,* J.Diff. Geom, **60** (2002) 199–293.

[Liu1] A. Liu, *Family blowup formula, admissible graphs and the enumeration of singular curves, I,* J. Differential Geometry **56** (2000), 381–379, MR1868322, Zbl 1036.14014.

[Liu2] A. Liu, *The algebraic proof of the universality theorem,* preprint, arXiv:math.AG/0402045.

[LM] M. Levine and F. Morel, *Algebraic cobordism,* Springer Monographs in Mathematics Springer, Berlin (2007).

[LP] M. Levine and R. Pandharipande, *Algebraic cobordism revisited,* Invent. Math. **276** (2009), No. 1, 63–130.

[LeeP] Y.-P. Lee and R. Pandharipande, *Algebraic cobordism of bundles on varieties,* preprint, arXiv:1002.1500, to appear in JEMS.

[LW] J. Li and B. Wu, *Good degeneration of Quot-schemes and coherent systems,* preprint, arXiv:1110.0390.

[Ran1] Z. Ran, *The degree of a Severi variety,* Bull. AMS. (New Series), **17** (1987), No. 1, 125–128.

[Ran2] Z. Ran, *Enumerative Geometry of Singular Plane Curves,* Invent. Math. **97** No. 3, (1989) 447–465.

[Tz] Y. Tzeng, *A proof to the Göttsche-Yau-Zaslow formula,* preprint arXiv:1009.5371, to appear in Journal of Diff. Geometry.

[Va] I. Vainsencher, *Enumeration of n-fold tangent hyperplanes to a surface,* Journal of Algebraic Geometry **4** (1995), 503–526.

[Vakil] R. Vakil, *Counting curves on rational surfaces,* Manuscripta Mathematica, **102** No. 1, (2000), 53–84.

[YZ] S. T. Yau and E. Zaslow, *BPS states, string duality, and nodal curves on K3,* Nuclear Physics B. **471** No. 3, (1996), 503–512.

Department of Mathematics, Harvard University, Cambridge, MA 02138, USA

E-mail address: ytzeng@math.harvard.edu

Current Developments in Mathematics, 2010

Some recent results on representations of p-adic special orthogonal groups

Jean-Loup Waldspurger

1. Motivation and results

1.1. Global motivation. Let V be a vector space over $\mathbb{Q}$, of finite dimension d, and let q be a non-degenerate quadratic form on V. We suppose given an orthogonal decomposition $V = D \oplus W$, where D is a line. Denote by G, resp. H, the special orthogonal group of V, resp. W. The group H is a subgroup of G, it is the subgroup of elements which act by identity on D. We assume, for instance, d even, and, to simplify, $d \geq 4$.

Denote by $\mathbb{A}$ the ring of adeles of $\mathbb{Q}$. The group $G(\mathbb{A})$ acts by right translations on the space $L^2(G(\mathbb{Q})\backslash G(\mathbb{A}))$ of square-integrable functions on $G(\mathbb{Q})\backslash G(\mathbb{A})$. Consider a closed subspace $\mathcal{V} \subset L^2(G(\mathbb{Q})\backslash G(\mathbb{A}))$, invariant by $G(\mathbb{A})$, such that the representation π of $G(\mathbb{A})$ on $\mathcal{V}$ is irreducible (there is no closed invariant subspace distinct from $\mathcal{V}$ and $\{0\}$). Such representation π is called automorphic of the discrete spectrum. The quotient $G(\mathbb{Q})\backslash G(\mathbb{A})$ is not compact in general, but we can define the notion of "rapid decay" for a function on $G(\mathbb{Q})\backslash G(\mathbb{A})$. Assume that the subset $\mathcal{V}^0$ of elements of rapid decay of $\mathcal{V}$ is dense in $\mathcal{V}$. Then we say that π is cuspidal. Let π be such automorphic cuspidal irreducible representation of $G(\mathbb{A})$ on a subspace $\mathcal{V} \subset L^2(G(\mathbb{Q})\backslash G(\mathbb{A}))$. Similarly, let σ be an automorphic cuspidal irreducible representation of $H(\mathbb{A})$ on a subspace $\mathcal{W} \subset L^2(H(\mathbb{Q})\backslash H(\mathbb{A}))$. For $\varphi \in \mathcal{V}^0$ and $\psi \in \mathcal{W}^0$, we define the integral

$$J(\psi, \varphi) = \int_{H(\mathbb{Q})\backslash H(\mathbb{A})} \overline{\psi(h)}\varphi(h)\, dh.$$

It is absolutely convergent and define a sesquilinear form J on $\mathcal{W}^0 \times \mathcal{V}^0$. A conjecture of Gross and Prasad ([GP], conjecture 14.8) says (in particular) that, if this sesquilinear form is non-zero, then a special value of some L-function is non-zero. What is this L-function? Of course, the notion of automorphic representation of the discrete spectrum is defined for all reductive groups over $\mathbb{Q}$ (some minor modifications are needed when the group has infinite center). Using a conjecture of Langlands, we can associate to π a representation $\boldsymbol{\pi}$ of the group $GL(d, \mathbb{A})$ that is not always an automorphic representation of the discrete spectrum, but is induced in a certain sense of such representation of a subgroup of $GL(d, \mathbb{A})$. We can consider $\boldsymbol{\pi}$ as a set $\{\boldsymbol{\pi}_1, \dots, \boldsymbol{\pi}_k\}$, where, for all $i = 1, \dots, k$, $\boldsymbol{\pi}_i$ is an automorphic representation of the discrete spectrum of $GL(d_i, \mathbb{A})$, with $\boldsymbol{\pi}_i \neq \boldsymbol{\pi}_{i'}$ if $i \neq i'$, and $d = d_1 + \cdots + d_k$. Similarly, we associate to σ a representation $\boldsymbol{\sigma}$ of $GL(d-2, \mathbb{A})$, that we can consider as a set $\{\boldsymbol{\sigma}_1, \dots, \boldsymbol{\sigma}_l\}$ (the group $GL(d-2)$ appears here for reasons that we will explain in 1.3). Jacquet, Piatetski-Shapiro and Shalika have defined a function $L(\boldsymbol{\pi} \times \boldsymbol{\sigma}, s)$ depending of a complex parameter s. The special value appearing in the Gross-Prasad conjecture is $L(\boldsymbol{\pi} \times \boldsymbol{\sigma}, 1/2)$. A more precise conjecture, due to Ichino and Ikeda ([II]), gives an exact formula expressing the square $|J(\psi, \varphi)|^2$ as the product of $L(\boldsymbol{\pi} \times \boldsymbol{\sigma}, 1/2)$ with more technical, but more elementary terms. Cf. 2.10 for a precise statement. When π and σ have arithmetical properties, such a formula implies rationality properties of the special value of the L-function.

A first problem is to understand when J is non-zero. The group $G(\mathbb{A})$ is a subgroup of $\prod_v G(\mathbb{Q}_v)$, with a suitable topology, where v describes the places of $\mathbb{Q}$, that is the set of prime numbers plus the infinite place $v = \infty$ for which $\mathbb{Q}_\infty = \mathbb{R}$. In a suitable sense, we can decompose $(\pi, \mathcal{V})$ as a tensor product $\otimes_v(\pi_v, \mathcal{V}_v)$ where, for all v, π_v is an irreducible unitary representation of $G(\mathbb{Q}_v)$ in an Hilbert space $\mathcal{V}_v$. Instead of $\mathcal{V}_v$, it is convenient to consider the dense subspace E_{π_v} of smooth vectors of $\mathcal{V}_v$. For $v = p$ a prime, E_{π_p} is the subspace of the elements $e \in \mathcal{V}_p$ such that the stabilizer $\{g \in G(\mathbb{Q}_p); \pi_p(g)e = e\}$ is open. This subspace is invariant by $G(\mathbb{Q}_p)$. Of course, the dimension of E_{π_p} is infinite in general. Similarly, we decompose $(\sigma, \mathcal{W})$ as a tensor product $\otimes_v(\sigma_v, \mathcal{W}_v)$. We see that, for all v, there exists a sesquilinear form J_v on $E_{\sigma_v} \times E_{\pi_v}$ such that $J = \prod_v J_v$ in a certain sense. Because of the integral definition of J, each form J_v satisfies the relation

$$(1) \qquad J_v(\sigma_v(h)e', \pi_v(h)e) = J_v(e', e)$$

for all $e \in E_{\pi_v}$, $e' \in E_{\sigma_v}$, $h \in H(\mathbb{Q}_v)$. Consider the space $Hom_{H(\mathbb{Q}_v)}(E_{\pi_v}, E_{\sigma_v})$. Its elements are the linear applications $l_v : E_{\pi_v} \to E_{\sigma_v}$ satisfying the relation $l_v(\pi_v(h)e) = \sigma_v(h)(l_v(e))$ for all $e \in E_{\pi_v}$ and $h \in H(\mathbb{Q}_v)$. To each such l_v, we can associate the sesquilinear form J^{l_v} on $E_{\sigma_v} \times E_{\pi_v}$ defined by

$$J^{l_v}(e', e) = (e', l_v(e))$$

for all $e' \in E_{\sigma_v}$ and $e \in E_{\pi_v}$, where the product of the right member is the hermitian product on $\mathcal{W}_v$. The form J^{l_v} satisfies the relation (1). We see easily that the application $l_v \mapsto J^{l_v}$ is a bijection between $Hom_{H(\mathbb{Q}_v)}(E_{\pi_v}, E_{\sigma_v})$ and the space of sesquilinear forms on $E_{\sigma_v} \times E_{\pi_v}$ satisfying (1). In particular, there exists $l_v \in Hom_{H(\mathbb{Q}_v)}(E_{\pi_v}, E_{\sigma_v})$ such that $J_v = J^{l_v}$. Of course, if $J \neq 0$, then $J_v \neq 0$ and $l_v \neq 0$. So a necessary condition such that $J \neq 0$ is that, for all v, the space $Hom_{H(\mathbb{Q}_v)}(E_{\pi_v}, E_{\sigma_v})$ is not zero.

The present formulation of the Ramanujan-Petersson conjecture says that if $\pi \simeq \otimes_v \pi_v$ is an irreducible automorphic and cuspidal representation of a group $GL(n, \mathbb{A})$, then all the components π_v are tempered. I will define later this notion. For our group G, this property is not true. The Arthur's conjecture says (among other things) what are the cuspidal automorphic representations π of $G(\mathbb{A})$ for which we can hope their components are tempered. The condition is that the elements $\boldsymbol{\pi}_i$ of $\boldsymbol{\pi} = \{\boldsymbol{\pi}_1, \ldots, \boldsymbol{\pi}_k\}$ are cuspidal automorphic representations, and not only automorphic representations of the discrete spectrum. In any case, the automorphic representations whose all components are (conjecturally) tempered are of special interest. So it's reasonable to restrict, at the present time, our local problem to tempered representations.

1.2. The local problem. Now, let F be $\mathbb{Q}_p$ for a prime p, or more generally, let F be a finite extension of $\mathbb{Q}_p$. Let $V = D \oplus W$, G and H as above except that the base field is now F and not $\mathbb{Q}$. Let π be a smooth irreducible representation of $G(F)$ in a complex space E_π and let σ be a smooth irreducible representation of $H(F)$ in a complex space E_σ. We consider the space $Hom_{H(F)}(E_\pi, E_\sigma)$. The first result is the following, due to Aizenbud, Gourevitch, Rallis and Schiffmann ([AGRS]).

THEOREM. $dim_{\mathbb{C}}(Hom_{H(F)}(E_\pi, E_\sigma)) \leq 1$.

We denote by $m(\pi, \sigma)$ this dimension. Introduce the contragredient $\check{\pi}$ of π. That is the transpose representation of π in the subspace $E_{\check{\pi}}$ of smooth vectors of the dual E_π^*. For $e \in E_\pi$ and $\check{e} \in E_{\check{\pi}}$, we define the coefficient $c_{e,\check{e}}(g) = <\pi(g)e, \check{e}>$ for $g \in G(F)$ ($<.,.>$ is the natural pairing between E_π and its dual). We say that π is tempered if, for all e and $\check{e}$, the function $c_{e,\check{e}}$ is of moderate growth, in a suitable sense. The same definition holds for σ. The central problem of my paper is to compute $m(\pi, \sigma)$ for π and σ tempered.

In fact, it's more convenient to consider two pairs (V_i, W_i) and (V_a, W_a) instead of one pair (V, W). We define the discriminant $\delta \in F^\times/F^{\times,2}$ of q (or of (V, q)) by $\delta = (-1)^{[d/2]}det(q)$, where the determinant is computed using any basis of V and $[d/2]$ is the integral part of $d/2$ (we have assumed d even in 1.1, but the same formula holds if d is odd). Recall that, for a given dimension and a given discriminant, there exists exactly two classes of equivalence of spaces provided with quadratic forms, of the given dimension and the given discriminant. We consider two such spaces V_i and V_a, with quadratic forms q_i and q_a, of the same even dimension d and the same discriminant δ, that are not equivalent. We assume given an orthogonal decomposition $V_i = D \oplus W_i$. Then we can decompose similarly $V_a = D \oplus W_a$ (with the same quadratic form on D). It's easy to prove that W_i and W_a, provided with the restrictions of q_i and q_a, have the same dimension and discriminant, but are not equivalent. We denote the four special orthogonal groups by G_i, G_a, H_i and H_a.

1.3. Parametrization of the tempered representations. Let $\bar{F}$ be an algebraic closure of F. Denote by W_F the Weil group of F. That is the subgroup of the elements of the Galois group $Gal(\bar{F}/F)$ that act by an entire power of the Frobenius on the residue field of $\bar{F}$ (the topology of W_F is finer that the induced topology). Let $W_{DF} = W_F \times SL(2, \mathbb{C})$ be the Weil-Deligne group. Assume again that the dimension $d = dim(V)$ is even. Then $d' = dim(W) = d - 1$ is odd. We introduce the symplectic group $Sp(d'-1, \mathbb{C}) \subset GL(d'-1, \mathbb{C})$. Consider the continuous homomorphisms $\varphi : W_{DF} \to Sp(d'-1, \mathbb{C})$ such that:

- the composed homomorphism
$$W_{DF} \xrightarrow{\varphi} Sp(d'-1, \mathbb{C}) \to GL(d'-1, \mathbb{C})$$
is a semi-simple representation of W_{DF};
- the restriction of φ to $SL(2, \mathbb{C})$ is algebraic.

The group $Sp(d'-1, \mathbb{C})$ acts by conjugacy on the set of such homomorphisms. Denote by $\Phi^{orth}(d')$ the set of conjugacy classes. Consider now the subset of φ as above, such that $\varphi(W_F)$ is a relatively compact subgroup of $Sp(d'-1, \mathbb{C})$. Denote by $\Phi^{orth}_{temp}(d')$ the set of conjugacy classes in this subset.

Denote by $\sharp$ one of the two indices i or a. The first assertion of a local conjecture of Langlands is that, to each $\varphi \in \Phi^{orth}_{temp}(d')$, is associated a finite set $\Pi^{H_\sharp}(\varphi)$ of smooth tempered irreducible representations of $H_\sharp(F)$, called an L-packet, and that the set of smooth tempered irreducible representations of $H_\sharp(F)$ is exactly the disjoint union of those L-packets $\Pi^{H_\sharp}(\varphi)$, when φ describes $\Phi^{orth}_{temp}(d')$.

Fix φ in $\Phi^{orth}_{temp}(d')$ (more exactly, φ is an homomorphism whose class is in this set). Denote by $S(\varphi)$ the centralizer of the image of φ in $Sp(d'-1, \mathbb{C})$, and by $S(\varphi)^0$ the identity component of $S(\varphi)$. The group $S(\varphi)/S(\varphi)^0$ is abelian, isomorphic to a finite product of copies of $\mathbb{Z}/2\mathbb{Z}$. Let $\Pi(\varphi)$ be the disjoint union of $\Pi^{H_i}(\varphi)$ and $\Pi^{H_a}(\varphi)$ and let $(S(\varphi)/S(\varphi)^0)^\wedge$ be the dual group of $S(\varphi)/S(\varphi)^0$. Then the conjecture says there is a bijection
$$\begin{array}{ccc} (S(\varphi)/S(\varphi)^0)^\wedge & \to & \Pi(\varphi) \\ \epsilon & \mapsto & \pi(\varphi, \epsilon). \end{array}$$
This bijection must satisfy some properties related to the theory of endoscopy and is uniquely determined by these properties.

The conjecture for the groups G_i and G_a is similar. The dimension d is even. The discriminant δ is an element of $F^\times/F^{\times 2}$, which defines a quadratic character of W_F, also denoted by δ. We introduce the complex orthogonal group $O(d, \mathbb{C})$ and we consider the homorphisms $\varphi : W_F \to O(d, \mathbb{C})$ satisfying the same conditions as above and the following. The sequence
$$W_F \to W_{DF} \xrightarrow{\varphi} O(d, \mathbb{C}) \xrightarrow{det} \{\pm 1\}$$
defines a quadratic character δ_φ. We assume $\delta_\varphi = \delta$. We denote $\Phi^{orth}(d, \delta)$ the set of conjugacy classes for the action of $SO(d, \mathbb{C})$ (and not $O(d, \mathbb{C})$) in

the set of such φ. We define as above the subset $\Phi^{orth}_{temp}(d,\delta)$. As previously, we conjecture that to each $\varphi \in \Phi^{orth}_{temp}(d,\delta)$ is associate a finite L-packet $\Pi^{G_\sharp}(\varphi)$ of smooth tempered irreducible representations of $G_\sharp(F)$ (for $\sharp = i$ or a). The set of smooth tempered irreducible representations of $G_\sharp(F)$ must be the disjoint union of the packets $\Pi^{G_\sharp}(\varphi)$ when φ describes $\Phi^{orth}_{temp}(d,\delta)$.

Fix $\varphi \in \Phi^{orth}_{temp}(d,\delta)$. Denote by $S(\varphi)$ the centralizer in $SO(d,\mathbb{C})$ of the image of φ, and by $S(\varphi)^0$ its identity component. The group $S(\varphi)/S(\varphi)^0$ is again a product of copies of $\mathbb{Z}/2\mathbb{Z}$. We denote by $\Pi(\varphi)$ the disjoint union of $\Pi^{G_i}(\varphi)$ and $\Pi^{G_a}(\varphi)$. Then the conjecture says there is a bijection

$$\begin{aligned}(S(\varphi)/S(\varphi)^0)^\wedge &\to \Pi(\varphi)\\ \epsilon &\mapsto \pi(\varphi,\epsilon).\end{aligned}$$

A great recent work of Arthur, but not yet published, proves a substantial part of the parametrization's conjectures, cf. 4.8 for more details.

1.4. Weak form of the Gross-Prasad conjecture. From now on, I assume the parametrization's conjectures as stated in Section 1.3. For a smooth irreducible representation π of $G_i(F)$, resp. $G_a(F)$, and for a smooth irreducible representation σ of $H_i(F)$, resp. $H_a(F)$, we define the multiplicity $m(\pi,\sigma) \in \{0,1\}$ as in Section 1.2.

THEOREM A. *Let $\varphi \in \Phi^{orth}_{temp}(d,\delta)$ and $\varphi^H \in \Phi^{orth}_{temp}(d'-1)$. There exists a unique pair $(\pi,\sigma) \in (\Pi^{G_i}(\varphi)\times\Pi^{H_i}(\varphi^H)) \sqcup (\Pi^{G_a}(\varphi)\times\Pi^{H_a}(\varphi^H))$ such that $m(\pi,\sigma)=1$.*

Cf. [W2], th. 7.10. It's the first assertion of the local Gross-Prasad conjecture ([GP], conjecture 10.7). This result was previously know in some special cases ([GR], [GGP1]).

1.5. Strong form of the Gross-Prasad conjecture. Let φ and φ^H as in the theorem A. This theorem defines a unique pair, which is parametrized by an element $(\epsilon,\epsilon^H) \in (S(\varphi)/S(\varphi)^0)^\wedge \times (S(\varphi^H)/S(\varphi^H)^0)^\wedge$. I state the following theorem in an imprecise form.

THEOREM B. *The parameters ϵ and ϵ^H can be explicitly computed.*

Cf. [W4], th. 4.9. This assertion is the second part of the local Gross-Prasad conjecture. As in the global situation, we can associate to φ a finite family $(\boldsymbol{\pi}_1,\dots,\boldsymbol{\pi}_k)$ of smooth irreducible representations of groups $GL(d_i,F)$, of the discrete serie, and similarly, we can associate to φ^H a finite family $(\boldsymbol{\sigma}_1,\dots,\boldsymbol{\sigma}_l)$ of representations of groups $GL(d'_j,F)$. Then ϵ and ϵ^H are related to the ϵ-factors $\epsilon(\boldsymbol{\pi}_i\times\boldsymbol{\sigma}_j,1/2,\psi_F)$ defined by Jacquet, Piatetski-Shapiro and Shalika.

1.6. Generalizations. Consider parameters $\varphi \in \Phi^{orth}(d, \delta)$ and $\varphi^H \in \Phi^{orth}(d'-1)$ that are not tempered. Using the theorem of Langlands' quotient, we can again define L-packets $\Pi(\varphi)$ and $\Pi(\varphi^H)$, cf. Section 6.1. In general, there is no conjecture (at the present time) that computes the multiplicities $m(\pi, \sigma)$. But the local Gross-Prasad conjecture holds if $\Pi(\varphi)$ and $\Pi(\varphi^H)$ are generic. We will define in 6.2 what is a generic L-packet. This property is weaker than to be tempered. For instance, all local components of a cuspidal automorphic representation of $GL(n, \mathbb{A})$ are generic.

THEOREM C (joint with Moeglin). *Let $\varphi \in \Phi^{orth}(d, \delta)$ and $\varphi^H \in \Phi^{orth}(d')$. Assume $\Pi(\varphi)$ and $\Pi(\varphi^H)$ are generic. Then the theorems A and B holds.*

Cf. [MW], introduction.

Instead of a decomposition $V = D \oplus W$, with a line D, assume now there is a decomposition $V = Z \oplus W$, where $dim(Z)$ is odd and the special orthogonal group of Z is split. For smooth irreducible representations π of $G(F)$ and σ of $H(F)$, we can again define a multiplicity $m(\pi, \sigma)$, using a variant of Whittaker's models. We have again $m(\pi, \sigma) = 0$ or 1. The theorems A, B, and C holds well in this case.

We have assumed that, in the pair $(dim(V), dim(W))$, the greatest d was even and the smallest was odd. But we can assume the greatest is odd and the smallest is even. The results are the same.

Our work concerns only special orthogonal groups. But Gan, Gross and Prasad consider other cases, cf. [GGP2]. If you are optimistic, you can hope that the same method holds in some other cases, for instance for unitary groups.

2. An integral formula

2.1. Multiplicities for more general situation. As before, let F be a finite extension of $\mathbb{Q}_p$. Let V be a vector space over F, of finite dimension d, and let q be a non-degenerate quadratic form on V. We remove the assumption that d is even. Assume given:

- an orthogonal decomposition $V = W \oplus Z$;
- a basis $\{v_i; i = -r, \dots, r\}$ of Z, for some $r \in \mathbb{N}$, and an element $\nu \in F^\times$ such that, if $v = \sum_{i=-r,\dots,r} x_i v_i \in Z$,

$$q(v,v)/2 = \nu(-1)^d x_0^2 + \sum_{i=1,\dots,r} x_i x_{-i}.$$

In particular, the dimension $2r+1$ of Z is odd. We denote by G and H the special orthogonal groups of V and W. Introduce the flag of isotropic subspaces

$$Fv_r \subset Fv_r \oplus Fv_{r-1} \subset \dots \subset Fv_r \oplus \dots \oplus Fv_1$$

of V, and the set P of elements of G that fix these subspaces. It is a parabolic subgroup. We denote by U its unipotent radical. Fix a non-trivial and continuous character ψ_F of F. Define a character ξ of $U(F)$ by

$$\xi(u) = \psi_F\left(\sum_{i=0,\dots,r-1} q(u(v_i), v_{-i-1})\right).$$

The group H is included in a Levi component of P and ξ is fixed by conjugacy by elements of $H(F)$.

Now, let π, resp. σ, be an irreducible smooth representation of $G(F)$, resp. $H(F)$, in a complex space E_π, resp. E_σ. Let $Hom_{H(F),\xi}(E_\pi, E_\sigma)$ be the space of linear applications $\varphi : E_\pi \to E_\sigma$ such that

$$\varphi(\pi(hu)e) = \xi(u)\sigma(h)\varphi(e)$$

for all $h \in H(F)$, $u \in U(F)$, $e \in E_\pi$. Denote by $m(\pi,\sigma)$ the dimension over $\mathbb{C}$ of $Hom_{H(F),\xi}(E_\pi, E_\sigma)$. This dimension does not depend of the choice of ψ_F nor of the basis $\{v_i; i = -r,\dots,r\}$.

REMARK. When Z is a line, the situation is the same as in Section 1.2. When $dim(W) \leq 1$, we have $H = \{1\}$, the representation σ is the trivial representation of $\{1\}$ in $E_\sigma = \mathbb{C}$ and the elements of $Hom_{H(F),\xi}(E_\pi, E_\sigma)$ are usual Whittaker's functionals on E_π.

Gan, Gross and Prasad have generalized the theorem of Aizenbud, Gourevitch, Rallis and Schiffmann to the present situation: $m(\pi,\sigma)$ is equal to 0 or 1 ([GGP2], corollary 20.4).

We fix π and σ for the whole Section 2.

2.2. Tempered representations. We recall here in more details the definition of tempered representations, for a general reductive group L defined over F. Let ρ be a smooth irreducible representation of $L(F)$ in a complex space E_ρ. We have defined in 1.2 the coefficient $c_{e,\check{e}}$ for $e \in E_\rho$ and $\check{e} \in E_{\check{\rho}}$. Denote by Z_L the center of L. We say that ρ is of the discrete serie if and only if

- the central character of ρ is unitary;
- for all e and $\check{e}$, the absolute value $|c_{e,\check{e}}|$ is square-integrable on $L(F)/Z_L(F)$ (by the first condition, it is a function on this quotient).

Now consider a parabolic subgroup Q of L and a Levi component M of Q, with Q and M defined over F. Let ρ^M be a smooth irreducible representation of $M(F)$, of the discrete serie. Harish-Chandra has defined the induced representation $Ind_Q^L(\rho^M)$ of $L(F)$. It is not irreducible in general, but semi-simple and of finite lenght. We say that the representation ρ of $L(F)$ is tempered if and only if there exists Q, M and ρ^M as above such that ρ is an irreducible component of $Ind_Q^L(\rho^M)$. In fact, the pair (M, ρ^M) is uniquely determined by ρ up to conjugacy by $L(F)$. It is not obvious, but true, that a representation is tempered if and only if its coefficients are of moderate growth, in some suitable sense.

2.3. Some results of Harish-Chandra. Let π as before. Denote by $C_c^\infty(G(F))$ the space of functions $f : G(F) \to \mathbb{C}$ that are locally constant and of compact support (I will use later similar notation for other groups). Fix an Haar mesure on $G(F)$. Then $C_c^\infty(G(F))$ acts on the space E_π by

$$\pi(f)e = \int_{G(F)} f(g)\pi(g)e\, dg,$$

for $f \in C_c^\infty(G(F))$ and $e \in E_\pi$. The operator $\pi(f)$ has finite rank, so we can define the distribution $f \mapsto trace(\pi(f))$. Harish-Chandra has proved that this distribution was locally integrable, i.e. there exists a locally integrable function χ_π on $G(F)$ such that

$$trace(\pi(f)) = \int_{G(F)} f(g)\chi_\pi(g)\, dg.$$

The function χ_π is invariant by conjugacy. Let t be a semi-simple element of $G(F)$. If t is strongly regular (i.e. the centralizer $Z_G(t)$ of t in G is a torus), χ_π is constant in a neighbourhood of t. It's not true if t is not strongly regular. In the general case, denote by $\mathfrak{g}_t$ the Lie algebra of the identity component $G_t = Z_G(t)^0$ of $Z_G(t)$. Denote by $Nil(\mathfrak{g}_t(F))$ the set of nilpotent orbits in $\mathfrak{g}_t(F)$, for the adjoint action of $Z_G(t)^0(F)$. Let $U \in Nil(\mathfrak{g}_t(F))$, fix a measure on U, invariant by this adjoint action. We define the nilpotent orbital integral

$$f \mapsto J_U(f) = \int_U f(u)\, du$$

for $f \in C_c^\infty(\mathfrak{g}_t(F))$. Fix a non-degenerate quadratic form $q_{\mathfrak{g}_t}$ on $\mathfrak{g}_t(F)$, invariant by the adjoint action of $G_t(F)$. We define a Fourier transform $f \mapsto \hat{f}$ on $C_c^\infty(\mathfrak{g}_t(F))$ by the usual formula

$$\hat{f}(X) = \int_{\mathfrak{g}_t(F)} \psi_F(q_{\mathfrak{g}_t}(X,Y))f(Y)dY.$$

Then we have the Fourier transform of the nilpotent orbital integral $f \mapsto J_U(\hat{f})$. Moreover, for $X \in \mathfrak{g}_t(F)$ near 0, we define as usual $exp(X) \in G_t(F)$. Harish-Chandra proves ([HC], theorem 16.2) that there exists an open neighbourhood ω of 0 in $\mathfrak{g}_t(F)$ and a unique family $(c_{\pi,U}(t))_{U\in Nil(\mathfrak{g}_t(F))}$ of complex numbers such that, for $f \in C_c^\infty(\omega)$, we have the identity

$$\int_\omega \chi_\pi(exp(X)t)f(X)\, dX = \sum_{U\in Nil(\mathfrak{g}_t(F))} c_{\pi,U}(t)J_U(\hat{f}).$$

Of course, we need to define precisely the measures: the measure on $\mathfrak{g}_t(F)$ is autodual for the Fourier transform; using the quadratic form $q_{\mathfrak{g}_t}$, $U(F)$ appears as a symplectic variety and the measure on $U(F)$ is deduced from this structure. We see that the coefficients $c_{\pi,U}(t)$ are independent of the choice of the quadratic form $q_{\mathfrak{g}_t}$.

2.4. Some tori. Consider the set $\underline{T}$ of subtori $T \subset H$, defined over F, not maximal in general, such that there exists an orthogonal decomposition $W = W' \oplus W''$ satisfying the following conditions:

(1) $dim(W')$ is even;
(2) denote by H' the special orthogonal group of W' (it's a subgroup of H); then $T \subset H'$ and T is a maximal subtorus of H';
(3) the maximal F-split subtorus of T is $\{1\}$;
(4) if d is odd, the special orthogonal group G'' of $W'' \oplus Z$ is split over F; if d is even, the special orthogonal group H'' of W'' is split over F.

Let $T \in \underline{T}$ and $t \in T(F)$. If t is in general position, $G_t = T \times G''$ and $Nil(\mathfrak{g}_t(F)) = Nil(\mathfrak{g}''(F))$, where $\mathfrak{g}''$ is of course the Lie algebra of G''. The coefficient $c_{\pi,U}(t)$ is defined for all $U \in Nil(\mathfrak{g}''(F))$. Assume d odd. Then G'' is a split special orthogonal group in odd dimension, and it's well know that there exists a unique regular nilpotent orbit in $\mathfrak{g}''(F)$. Denote by U_{reg} this orbit and let $c_\pi(t) = c_{\pi,U_{reg}}(t)$. Assume now d even. Then there exists several regular nilpotent orbits in $\mathfrak{g}''(F)$. But consider the subspace $Z_0 \subset Z$ generated by the vectors v_i for $i \neq 0$. It is a sum of hyperbolic planes. Consider the special orthogonal group H_0'' of $W'' \oplus Z_0$. Because H'' is split, H_0'' is also split. For the same reason as before, there exists a unique regular nilpotent orbit in $\mathfrak{h}_0''(F)$. By the inclusion $H_0'' \subset G''$, it is included in some nilpotent orbit of $\mathfrak{g}''(F)$ and in fact in a regular nilpotent orbit. We denote by U_{reg} this orbit and we define $c_\pi(t) = c_{\pi,U_{reg}}(t)$ as before. Then c_π is a function defined almost everywhere on $T(F)$.

Similarly, we have $H_t = T \times H''$ and $Nil(\mathfrak{h}_t(F)) = Nil(\mathfrak{h}''(F))$. If d is even, H'' is split and in "odd dimension", so $\mathfrak{h}''(F)$ has a unique nilpotent regular orbit U_{reg}^H. If d is odd, and $dim(W'') \leq 2$, $\mathfrak{h}''(F)$ has again a unique nilpotent regular orbit U_{reg}^H. Assume d odd and $dim(W'') \geq 4$. Because G'' is split, we see that there exists an orthogonal decomposition $W'' = W_0 \oplus D_0$, where D_0 is a line and the space $D_0 \oplus Fv_0$ is an hyperbolic plane. As above, the Lie algebra of the special orthogonal group of W_0 has a unique regular nilpotent orbit. This orbit is included in a nilpotent orbit of $\mathfrak{h}''(F)$, denote by U_{reg}^H this orbit. In all cases, let $c_\sigma(t) = c_{\sigma,U_{reg}^H}(t)$. Then c_σ is a function defined almost everywhere on $T(F)$.

We define two others functions almost everywhere on $T(F)$. Let $t \in T(F)$ in general position. We put

$$\Delta(t) = |det((1-t)_{|W'})|_F,$$

where $|.|_F$ is the usual absolute value of F;

$$D^H(t) = |det((ad(t)-1)_{|\mathfrak{h}(F)/\mathfrak{h}_t(F)})|_F.$$

And we denote by $W(H,T)$ the quotient of the normalizer of T in $H(F)$ by the centralizer. It is a finite group.

2.5. The integral formula. The group $H(F)$ acts by conjugacy on $\underline{\mathcal{T}}$. We fix a set $\mathcal{T}$ of representatives of conjugacy classes. Let

$$m_{geom}(\pi,\sigma) = \sum_{T\in\mathcal{T}} |W(H,T)|^{-1} \int_{T(F)} c_\pi(t)c_\sigma(t)D^H(t)\Delta(t)^r\,dt,$$

where the measures on the tori have total mass 1.

PROPOSITION. *(i) The above formula is absolutely convergent.* *(ii) Assume π and σ are tempered. Then $m(\pi,\sigma) = m_{geom}(\pi,\sigma)$.*

Cf. [W2], théorème 7.1.

2.6. About the proof: the geometric side. The proof of the proposition is very long. I will describe the starting point only. We choose a family $(K_N)_{N\geq 1}$ of compact open subsets of $H(F)U(F)\backslash G(F)$ such that $K_N \subset K_{N+1}$ and $\bigcup_{N\geq 1} K_N = H(F)U(F)\backslash G(F)$. Let κ_N be the characteristic function of the inverse image of K_N in $G(F)$. For $N \geq 1$ and $f \in C_c^\infty(G(F))$, define

$$J_N(\theta_\sigma, f) = \int_{H(F)U(F)\backslash G(F)} \times \int_{H(F)}\int_{U(F)} \theta_\sigma(h)f(g^{-1}hug)\xi(u)^{-1}\,du\,dh\,\kappa_N(g)\,dg.$$

This expression is absolutely convergent. We want to compute its limit when N tends to ∞.

Here, the idea is the same as for the Arthur's local trace formula. There are two ways for the computation. The first is "geometric": we express the integral using orbital integrals or weighted orbital integrals of the function f. In fact, it is more complicated as for the local trace formula, because the integrals have more singularities as in this case. These singularities are reflected in the final formula by the presence of integrals over non maximal tori: these integrals appear as singularities of integrals over maximal tori. And its not possible to compute the limit of $J_N(\theta_\sigma, f)$ for all choice of the sets K_N and all functions f: we must choose particular K_N and we must assume that f is strongly cuspidal. That means that for all proper parabolic subgroup $P' = M'U'$ of G, defined over F, where M' is a Levi component of P' and U' is the unipotent radical of P', we have

$$\int_{U'(F)} f(m'u')\,du' = 0$$

for all $m' \in M'(F)$. Under those conditions, we can compute the limit as

$$lim_{N\to\infty} J_N(\theta_\sigma, f) = m_{geom}(f, \sigma)$$

where $m_{geom}(f, \sigma)$ is similar to $m_{geom}(\pi, \sigma)$. The function c_π is replaced by a function c_f deduced from weighted orbital integrals of f. Cf. [W1] théorème 7.8.

I will give a basic example of the computation. Consider the case where $dim(V) = 3$, $dim(W) = 2$ and G and H are split. We can identify G with $PGL(2)$ and H with its diagonal subtorus. Let $K = PGL(2, \mathfrak{o}_F)$ be the standard maximal subgroup of $G(F)$, where $\mathfrak{o}_F$ is the ring of integers of F. We can write every element $g \in G(F)$ as a product

$$g = \begin{pmatrix} a & 0 \\ 0 & 1 \end{pmatrix} \begin{pmatrix} 1 & b \\ 0 & 1 \end{pmatrix} k,$$

with $a \in F^\times$, $b \in F$ and $k \in K$ (we consider the two first matrices as elements of $PGL(2, F)$). The measure dg decomposes as $dg = |a|_F^{-1} da\, db\, dk$. We assume that $\kappa_N(g) = 1$ if $val_F(b) \geq -N$ and $\kappa_N(g) = 0$ if $val_F(b) < -N$, where val_F is the usual valuation of F. We can consider the character θ_σ as an usual character of $F^\times$. Then

$$J_N(\theta_\sigma, f) = \int_K \int_{F^\times} \int_{b\in F; val_F(b)\geq -N} \theta_\sigma(a)$$
$$f\left(k^{-1} \begin{pmatrix} 1 & -b \\ 0 & 1 \end{pmatrix} \begin{pmatrix} a & 0 \\ 0 & 1 \end{pmatrix} \begin{pmatrix} 1 & b \\ 0 & 1 \end{pmatrix} k\right) db\, |a|_F^{-1} da\, dk.$$

We define f^K by

$$f^K(g) = \int_K f(k^{-1} g k) dk.$$

Then

$$(1) \qquad J_N(\theta_\sigma, f) = \int_{F^\times} \int_{b\in F; val_F(b)\geq -N} \theta_\sigma(a) f^K$$
$$\times \left(\begin{pmatrix} a & (a-1)b \\ 0 & 1 \end{pmatrix}\right) db\, |a|_F^{-1} da.$$

What happens if we suppress the condition $val_F(b) \geq -N$? Changing b in $(a-1)^{-1} b$, we obtain the integral

$$(2) \qquad \int_{F^\times} \int_F \theta_\sigma(a) f^K \left(\begin{pmatrix} a & b \\ 0 & 1 \end{pmatrix}\right) db\, |a|_F^{-1} |a-1|_F^{-1} da.$$

The support of the integrand is compact. But the integral is not convergent because the function $a \mapsto |a-1|_F^{-1}$ is not integrable in a neighbourhood of $a = 1$ (the point $a = 1$ corresponds to the point $1 \in H(F)$). So we choose some auxiliary constant $\epsilon > 0$ and we decompose our integral (1) in the sum of an integral over the domain $|a-1|_F > \epsilon$ and an integral over the domain $|a-1|_F \leq \epsilon$. In the first integral, we can use the dominate convergence theorem: its limit when $N \to \infty$ is the integral (2), restricted to the domain

$|a-1|_F > \epsilon$. In fact, the limit is 0: the integral over b is 0 because f is strongly cuspidal. Consider the second integral. Because f^K is locally constant of compact support, we can choose ϵ such that

$$f^K\left(\begin{pmatrix} a & (a-1)b \\ 0 & 1 \end{pmatrix}\right) = f^K\left(\begin{pmatrix} 1 & (a-1)b \\ 0 & 1 \end{pmatrix}\right)$$

for every b and every a such that $|a-1|_F < \epsilon$. We can also assume that $\theta_\sigma(a) = 1$ for every such a. Let $x = a-1$. The condition $|a-1|_F < \epsilon$ is equivalent to $val_F(x) > n$, where n is some big integer determined by ϵ. Then our integral is equal to

$$\int_{x\in F; val_F(x)>n} \int_{b\in F; val_F(b)\geq -N} f^K\left(\begin{pmatrix} 1 & xb \\ 0 & 1 \end{pmatrix}\right) db\, dx.$$

We change b to bx^{-1} and we obtain

$$\int_{x\in F; val_F(x)>n} \int_{b\in F; val_F(b)\geq -N+val_F(x)} f^K\left(\begin{pmatrix} 1 & b \\ 0 & 1 \end{pmatrix}\right) db\, |x|_F^{-1} dx.$$

Assume that we can permute the two integrals. Then we obtain

$$\int_{b\in F} \int_{x\in F; n<val_F(x)\leq N+val_F(b)} f^K\left(\begin{pmatrix} 1 & b \\ 0 & 1 \end{pmatrix}\right) |x|_F^{-1} dx\, db.$$

Because the support of f^K is compact, we can assume $val_F(b) \geq c$ for some constant c. Then $N + val_F(b) \geq n$ if N is big and the inner integral is equal to $N + val_F(b) - n$ (assuming that the measure of the set of units is 1). We obtain

$$(3) \qquad \int_{b\in F} f^K\left(\begin{pmatrix} 1 & b \\ 0 & 1 \end{pmatrix}\right) (N - n + val_F(b))\, db.$$

This expression is absolutely convergent. If we replace f^K by its absolute value in the preceding computation, we obtain a similar result. Then the permutation of integrals is justified and our integral is equal to (3), or to the sum of

$$(4) \qquad (N-n) \int_{b\in F} f^K\left(\begin{pmatrix} 1 & b \\ 0 & 1 \end{pmatrix}\right) db$$

and

$$(5) \qquad \int_{b\in F} f^K\left(\begin{pmatrix} 1 & b \\ 0 & 1 \end{pmatrix}\right) val_F(b)\, db.$$

But (4) is 0 for the same reason as before: the function f is strongly cuspidal. Up to a constant, the expression (5) is the unipotent weighted orbital integral of f as defined by Arthur. Coming back to our general formula for $m_{geom}(f,\sigma)$, we have only one torus in $\mathcal{T}$: the torus $T = \{1\}$. For the unique element $1 \in T(F)$, we have $c_\sigma(1) = 1$ and $c_f(1)$ is the weighted orbital integral above.

2.7. The spectral side. The second way for computing the limit $lim_{N\to\infty} J_N(\theta_\sigma, f)$ is "spectral". We need to introduce some notations. We call "Levi subgroup of G" a subgroup $L \subset G$ defined over F that is a Levi component of some parabolic subgroup of G defined over F. We fix some minimal Levi subgroup M_0 of G and we denote by $\mathcal{L}(M_0)$ the set of Levi subgroup of G that contains M_0. Let $L \in \mathcal{L}(M_0)$. Denote by $X^*(L)$ the group of algebraic characters of L defined over F. Let $\mathcal{A}_L^* = X^*(L) \otimes_{\mathbb{Z}} \mathbb{R}$, $\mathcal{A}_{L,\mathbb{C}}^* = \mathcal{A}_L^* \otimes_{\mathbb{R}} \mathbb{C}$ and $a_L = dim(\mathcal{A}_{L,\mathbb{C}}^*)$. Each element $\lambda \in \mathcal{A}_{L,\mathbb{C}}^*$ defines a continuous homomorphism $\lambda_L : L(F) \to \mathbb{C}^\times$ as follows. If $\lambda = \sum_{i=1,\ldots,n} s_i \chi_i$, with $s_i \in \mathbb{C}$ and $\chi_i \in X^*(L)$, and if $l \in L(F)$, we put

$$\lambda_L(l) = \prod_{i=1,\ldots,n} |\chi_i(l)|_F^{s_i}.$$

Let's remark that the values $|\chi_i(l)|_F$ are all integral powers of the number of elements q of the residual field of F. This implies that $\lambda_L = 1$ if $s_i \in (2\pi i/log(q))\mathbb{Z}$ for all i. So the homomorphism λ_L depends only of the image of λ in $\mathcal{A}_{L,\mathbb{C}}^*/i\mathcal{A}_{L,F}^\vee$, where $\mathcal{A}_{L,F}^\vee$ is some lattice in $\mathcal{A}_L^*$. If ρ is a smooth representation of $L(F)$, we denote by ρ_λ the representation $l \mapsto \lambda_L(l)\rho(l)$. If ρ is tempered, or of the discrete serie, and if $\lambda \in i\mathcal{A}_L^*$, then ρ_λ is also tempered, or of the discrete serie. We denote by $\Pi_2(L)$ the set of irreducible representations of $L(F)$ of the discrete serie. Arthur has defined a set $\Pi_{ell}(L)$ of "elliptic" representations that are not always irreducible. It is slightly bigger as $\Pi_2(L)$. We can say that it is the set of virtual representations that satisfy the same condition of growth as the irreducible representations of the discrete serie. For a smooth irreducible representation ρ of $L(F)$, of finite length, Arthur has defined its weighted character, that is a distribution $f \mapsto J_L^G(\rho, f)$ on $C_c^\infty(G(F))$. When $L = G$, it is the usual character θ_ρ and it is an invariant distribution (invariant by conjugacy by $G(F)$). But, if $L \subsetneq G$, the definition involves intertwining operators and the distribution is not invariant.

A Levi subgroup $L \in \mathcal{L}(M_0)$ is not a special orthogonal group, if $L \neq G$. But it is a product

$$L = L_0 \times GL(n_1) \times \cdots \times GL(n_k).$$

The group L_0 is the special orthogonal group of a subspace $V_0 \subset V$, of which the orthogonal is a sum of hyperbolic planes. Let ρ be an smooth irreducible representation of $L(F)$. We write $\rho = \rho_0 \otimes (\otimes_i \rho_i)$, where ρ_0 is a representation of $L_0(F)$ and the ρ_i are representations of the factors $GL(n_i)$. Up to conjugacy, the spaces V_0 and W satisfy the same conditions as our starting spaces V and W (may be the roles of V and W are reversed). So we can define the multiplicity $m(\rho_0, \sigma)$ (or $m(\sigma, \rho_0)$ if the roles of V and W are reversed) and we put $m(\rho, \sigma) = m(\rho_0, \sigma)$ (or $m(\sigma, \rho_0)$). Here the representation ρ was irreducible, but it is possible to adapt the definition to the case of an elliptic non irreducible representation.

We come back to our expression $J_N(\theta_\sigma, f)$. Using the Plancherel formula, due to Harish-Chandra in this case, we express f as an integral over the set of tempered irreducible representations of $G(F)$ of the traces of f in these representations. After a long computation, which follows closely the one of Arthur (but it's also necessary to assume that f is strongly cuspidal), we obtain

$$lim_{N\to\infty} J_N(\theta_\sigma, f) = \sum_{L\in\mathcal{L}(M_0)} |W^L||W^G|^{-1}(-1)^{a_L} \sum_{\rho\in\Pi_{ell}(L)/i\mathcal{A}_L^*; m(\rho,\sigma)=1}$$
$$c(\rho)\int_{i\mathcal{A}_L^*/i\mathcal{A}_{L,F}^\vee} J_L^G(\rho_\lambda, f)\, d\lambda.$$

Cf. [W2] théorème 6.1. The terms W^L and W^G are the usual Weyl groups of L and G, relative to the minimal Levi M_0. The term $c(\rho)$ is some explicit constant, depending of the measures. The important point is the condition $m(\rho, \sigma) = 1$ that appears in our formula. We explain this fact in 2.9.

Using the method of Arthur, we can transform the preceding formulas such that they contain only invariant distributions (and not weighted characters, that are not invariant). Then, assuming π of the discrete serie, we choose for f a pseudo-coefficient of π, that is a function such that $\chi_\pi(f) = 1$ but $\chi_{\pi'}(f) = 0$ for all irreducible tempered representation π' of $G(F)$, $\pi' \neq \pi$. Then we obtain the second assertion of the proposition 2.5 by comparison of the two limits computed in 2.6 and above. The general case where π is only tempered can be deduced by studying the behaviour of multiplicities by induction.

2.8. Construction of intertwining homomorphisms. A technical, but important point, is to describe explicitly the elements of the space $Hom_{H(F),\xi}(E_\pi, E_\sigma)$ (that is a complex line, or $\{0\}$). Assume for simplicity $r = 0$, i.e. $dim(V) = dim(W) + 1$. Then $Hom_{H(F),\xi}(E_\pi, E_\sigma) = Hom_{H(F)}(E_\pi, E_\sigma)$. For $e \in E_\pi$ and $\check{e} \in E_{\check{\pi}}$, define a coefficient

$$g \mapsto < \check{e}, \pi(g)e >$$

on $G(F)$. For $e' \in E_\sigma$ and $\check{e}' \in E_{\check{\sigma}}$, define a coefficient

$$h \mapsto < \check{\sigma}(h)\check{e}', e' >$$

on $H(F)$. Consider the integral

$$L_{\pi,\sigma}(\check{e}, e, \check{e}', e') = \int_{H(F)} < \check{e}, \pi(h)e >< \check{\sigma}(h)\check{e}', e' > dh.$$

Assume π and σ are tempered. Then this integral is absolutely convergent. It defines a linear form $L_{\pi,\sigma}$ on $E_{\check{\pi}} \otimes E_\pi \otimes E_{\check{\sigma}} \otimes E_\sigma$. Assume $L_{\pi,\sigma} \neq 0$. Fix $\check{e}_0$, e_0, $\check{e}'_0$ and e'_0 such that $L_{\pi,\sigma}(\check{e}_0, e_0, \check{e}'_0, e'_0) \neq 0$. We define a linear application $l : E_\pi \to E_{\check{\sigma}}^*$ by the formula

$$< \check{e}', l(e) >= L_{\pi,\sigma}(\check{e}_0, e, \check{e}', e'_0)$$

for $e \in E_\pi$ and $\check{e}' \in E_{\check{\sigma}}$. It's easy to prove that l takes values in $E_\sigma \subset E_{\check{\sigma}}^*$ and $l \in Hom_{H(F)}(E_\pi, E_\sigma)$. Because $L_{\pi,\sigma} \neq 0$, we have also $l \neq 0$. Therefore, $Hom_{H(F)}(E_\pi, E_\sigma) \neq \{0\}$ if $L_{\pi,\sigma} \neq 0$. The converse is more difficult, but true, cf. [W2] proposition 5.7.

PROPOSITION. *Assume π and σ tempered. Then $L_{\pi,\sigma} \neq 0$ if and only if $Hom_{H(F)}(E_\pi, E_\sigma) \neq \{0\}$.*

Variant. When π and σ are unitary, in particular when they are tempered, we can fix positive definite hermitian products over the spaces E_π and E_σ (linear in the second variable and anti-linear in the first) invariant by the actions of $G(F)$ and $H(F)$, and define

$$\mathcal{L}_{\pi,\sigma}(e_1, e_2, e_1', e_2') = \int_{H(F)} (e_1, \pi(h)e_2)(\sigma(h)e_1', e_2')\, dh$$

for $e_1, e_2 \in E_\pi$ and $e_1', e_2' \in E_\sigma$. We have a similar result for this semi-linear form. Moreover, if we assume $Hom_{H(F)}(E_\pi, E_\sigma) \neq \{0\}$ and if we fix some non-zero element l in this space, we see that there exists some constant $c \neq 0$ such that

$$(1) \qquad \mathcal{L}_{\pi,\sigma}(e_1, e_2, e_1', e_2') = c(e_1', l(e_2))(l(e_1), e_2').$$

2.9. The basic computation of the spectral side. Here we assume that Z is a line as in the Section 1. The Plancherel formula express f as an integral over the set of tempered irreducible representations of $G(F)$ of the traces of f in these representations. Consider an irreducible representation ρ of $G(F)$ of the discrete serie. The contribution of ρ to this formula is the function

$$g \mapsto d(\rho) trace(\rho(g^{-1})\rho(f)),$$

where $d(\rho)$ is the formal degree of ρ as defined later. Assume that it is the only contribution to the Plancherel formula. So $f(g)$ is equal to the expression above. Then

$$\begin{aligned} &J_N(\theta_\sigma, f) \\ (1) \qquad &= d(\rho) \int_{H(F)\backslash G(F)} \int_{H(F)} \theta_\sigma(h) trace(\rho(g^{-1}h^{-1}g)\rho(f)) dh\, \kappa_N(g)\, dg. \end{aligned}$$

Fix invariant positive definite hermitian products over the spaces E_ρ and E_σ and fix orthonormal basis $(e_i)_{i\in I}$ of E_ρ and $(e_j')_{j\in J}$ of E_σ. Fix $g \in G(F)$. For $h \in H(F)$, we have

$$\begin{aligned} trace(\rho(g^{-1}h^{-1}g)\rho(f)) &= \sum_{i\in I} (e_i, \rho(g^{-1}h^{-1}g)\rho(f)e_i) \\ &= \sum_{i\in I} (\rho(g)e_i, \rho(h^{-1}g)\rho(f)e_i). \end{aligned}$$

Because f is biinvariant by some open compact subgroup, we can restrict the sum to a finite subset $I_f \subset I$ independent of g. By definition of the character θ_σ, the inner integral of the expression (1) is equal to the trace of the operator

$$\int_{H(F)} trace(\rho(g^{-1}h^{-1}g)\rho(f))\sigma(h)\,dh,$$

that is

$$(2)\qquad \sum_{j\in J}(e'_j, \left(\int_{H(F)}\sum_{i\in I_f}(\rho(g)e_i,\rho(h^{-1}g)\rho(f)e_i)\,\sigma(h)dh\right)e'_j).$$

Fix $j\in J$. The corresponding term above is absolutely convergent and equal to

$$(3)\qquad \sum_{i\in I_f}\int_{H(F)}(\rho(g)e_i,\rho(h^{-1}g)\rho(f)e_i)(\sigma(h^{-1})e'_j,e'_j)\,dh.$$

Changing h to h^{-1}, the integral above is $\mathcal{L}_{\rho,\sigma}(\rho(g)e_i,\rho(g)\rho(f)e_i,e'_j,e'_j)$. If $m(\rho,\sigma)=0$, the semi-linear form $\mathcal{L}_{\rho,\sigma}$ is 0 by the proposition 2.8. Our computation shows that $J_N(\theta_\sigma,f)=0$. Assume now that $m(\rho,\sigma)=1$. Fix some non-zero element $l\in Hom_{H(F)}(E_\rho,E_\sigma)$ and let c be the non-zero constant such that 2.8(1) is satisfied. Then (3) is equal to

$$\sum_{i\in I_f}c(l(\rho(g)e_i),e'_j)(e'_j,l(\rho(g)\rho(f)e_i))$$

and (2) is equal to

$$\sum_{j\in J}\sum_{i\in I_f}c(l(\rho(g)e_i),e'_j)(e'_j,l(\rho(g)\rho(f)e_i)).$$

Using the equality

$$l(\rho(g)\rho(f)e_i)=\sum_{j\in J}(e'_j,l(\rho(g)\rho(f)e_i))e'_j,$$

we obtain that (2) is equal to

$$\sum_{i\in I_f}c(l(\rho(g)e_i),l(\rho(g)\rho(f)e_i)).$$

Fix $e_*\in E_\rho$ and $e'_*\in E_\sigma$ such that $(e'_*,l(e_*))=1$. We can multiply the expression above by $1=(e'_*,l(e_*))(l(e_*),e'_*)$. Using 2.8(1), we have

$$c(l(\rho(g)e_i),l(\rho(g)\rho(f)e_i))(e'_*,l(e_*))=\mathcal{L}_{\rho,\sigma}(\rho(g)e_i,e_*,e'_*,l(\rho(g)\rho(f)e_i))$$

Using the definition of this term, (2) becomes equal to

$$\sum_{i\in I_f}(l(e_*),e'_*)\int_{H(F)}(\sigma(h)e'_*,l(\rho(g)\rho(f)e_i))(\rho(g)e_i,\rho(h)e_*)\,dh.$$

Using once more 2.8(1), we have

$$(l(e_*), e'_*)(\sigma(h)e'_*, l(\rho(g)\rho(f)e_i)) = c^{-1}\mathcal{L}_{\rho,\sigma}(e_*, \rho(g)\rho(f)e_i, \sigma(h)e'_*, e'_*).$$

Using the definition of this term, (2) becomes equal to

$$c^{-1}\sum_{i\in I_f}\int_{H(F)}(\rho(g)e_i, \rho(h)e_*)\int_{H(F)}(\sigma(h'h)e'_*, e'_*)(e_*, \rho(h'g)\rho(f)e_i)\,dh'\,dh.$$

Coming back to (1), we obtain

$$J_N(\theta_\sigma, f) = d(\rho)c^{-1}\int_{H(F)\backslash G(F)}\sum_{i\in I_f}\int_{H(F)}(\rho(g)e_i, \rho(h)e_*)$$
$$\times\int_{H(F)}(\sigma(h'h)e'_*, e'_*)(e_*, \rho(h'g)\rho(f)e_i)\,dh'\,dh\,\kappa_N(g)\,dg.$$

Because ρ is of the discrete serie, a tedious computation shows that this expression is absolutely convergent, even if we suppress the term $\kappa_N(g)$. We obtain first that $lim_{N\to\infty}J_N(\theta_\sigma, f)$ is the same expression where $\kappa_N(g)$ is removed. Next, we can change h to $(h')^{-1}h$ and the double integration over $g\in H(F)\backslash G(F)$ and $h'\in H(F)$ to a unique integration over $g\in G(F)$. So

$$lim_{N\to\infty}J_N(\theta_\sigma, f) = d(\rho)c^{-1}\sum_{i\in I_f}\int_{G(F)}\int_{H(F)}(\rho(g)e_i, \rho(h)e_*)$$
$$\times(\sigma(h)e'_*, e'_*)(e_*, \rho(g)\rho(f)e_i)\,dh\,dg$$
$$(4)\qquad = c^{-1}\sum_{i\in I_f}\int_{H(F)}(\sigma(h)e'_*, e'_*)d(\rho)$$
$$\times\int_{G(F)}(\rho(g)e_i, \rho(h)e_*)(e_*, \rho(g)\rho(f)e_i)\,dg\,dh.$$

For every $\epsilon_1, \epsilon_2, \epsilon_3, \epsilon_4\in E_\rho$, we have the equality

$$d(\rho)\int_{G(F)}(\rho(g)\epsilon_1, \epsilon_2)(\epsilon_3, \rho(g)\epsilon_4)dg = (\epsilon_1, \epsilon_4)(\epsilon_3, \epsilon_2).$$

In fact, that is the definition of the formal degree of a representation of the discrete serie. Using this relation, we compute the inner integral of the expression (4), and we obtain

$$lim_{N\to\infty}J_N(\theta_\sigma, f) = c^{-1}\sum_{i\in I_f}\int_{H(F)}(\sigma(h)e'_*, e'_*)(e_i, \rho(f)e_i)(e_*, \rho(h)e_*)\,dh$$
$$= \sum_{i\in I_f}(e_i, \rho(f)e_i))c^{-1}\int_{H(F)}(\sigma(h)e'_*, e'_*)(e_*, \rho(h)e_*)\,dh.$$

The integral is again equal to $\mathcal{L}_{\rho,\sigma}(e_*, e_*, e'_*, e'_*)$. Using 2.8(1) and the definition of e_* and e'_*, it is equal to c. Then

$$lim_{N\to\infty} J_N(\theta_\sigma, f) = \sum_{i\in I_f} (e_i, \rho(f)e_i)) = trace(\rho(f)).$$

So we obtain the final formula

$$lim_{N\to\infty} J_N(\theta_\sigma, f) = \begin{cases} 0, & \text{if } m(\rho,\sigma) = 0, \\ trace(\rho(f)), & \text{if } m(\rho,\sigma) = 1, \end{cases}$$

in others words

$$lim_{N\to\infty} J_N(\theta_\sigma, f) = m(\rho,\sigma) trace(\rho(f)).$$

It is the formula of 2.7 for our choice of f.

2.10. Intermission: the conjecture of Ichino and Ikeda. We interrupt our work to give the precise statement of the conjecture of Ichino and Ikeda quoted in 1.1. It shows the role of the semi-linear form introduced in Section 2.8. For this section, the ground field is $\mathbb{Q}$ again and we assume $d = dim(V) = dim(W) + 1$ and d is even. Let $\mathcal{V}$, resp. $\mathcal{W}$, be an irreducible subspace of the space of cuspidal automorphic forms on $G(\mathbb{Q})\backslash G(\mathbb{A})$, resp. $H(\mathbb{Q})\backslash H(\mathbb{A})$. Denote by π, resp. σ, the representation of $G(\mathbb{A})$ on $\mathcal{V}$, resp. of $H(\mathbb{A})$ on $\mathcal{W}$. We decompose π and σ as tensor products, in certain sense, of representations π_v and σ_v, where v describes the places of $\mathbb{Q}$. For every v, we denote by E_{π_v} and E_{σ_v} the space of smooth vectors of π_v and σ_v (if $v = \infty$ is the real place, we add the condition that the elements of E_{π_∞} and E_{σ_∞} are K-finite, a condition that we don't explain). Let $\varphi \in \mathcal{V}$ such that $\varphi \simeq \otimes_v e_v$, where $e_v \in E_{\pi_v}$ for every v, and similarly let $\psi \in \mathcal{W}$ such that $\psi \simeq \otimes e'_v$. Fix a finite set S of places of $\mathbb{Q}$, containing ∞, such that all the data are unramified for every $v \notin S$. That means that for $v \notin S$, we can fix maximal compact subgroups of $G(\mathbb{Q}_v)$ and $H(\mathbb{Q}_v)$ that are hyperspecial in the terminology of Bruhat and Tits and such that e_v and e'_v are invariant by these subgroups.

We assume that all local components π_v and σ_v of π and σ are tempered. We associate to π and σ sets $\{\boldsymbol{\pi}_1, \dots, \boldsymbol{\pi}_k\}$ and $\{\boldsymbol{\sigma}_1, \dots, \boldsymbol{\sigma}_l\}$ as in 1.1. Their elements are unitary cuspidal automorphic representations of some groups, say $GL(d_i, \mathbb{A})$ for $\boldsymbol{\pi}_i$ and $GL(d'_j, \mathbb{A})$ for $\boldsymbol{\sigma}_j$. For two such representations $\boldsymbol{\tau}$ of $GL(N, \mathbb{A})$ and $\boldsymbol{\tau}'$ of $GL(N', \mathbb{A})$, unramified outside S, we consider the partial L-function $L^S(\boldsymbol{\tau} \times \boldsymbol{\tau}', s)$, depending of a complex parameter s. For $Re(s) \gg 0$, it is defined by an eulerian product

$$L^S(\boldsymbol{\tau} \times \boldsymbol{\tau}', s) = \prod_{p\notin S} L_p(\boldsymbol{\tau} \times \boldsymbol{\tau}', s).$$

For $p \notin S$, the theory of Hecke operators associates to π_p and σ_p complex parameters $\lambda_{1,p}, \dots, \lambda_{N,p}$ and $\mu_{1,p}, \dots, \mu_{N',p}$ and we have

$$L_p(\boldsymbol{\tau} \times \boldsymbol{\tau}', s) = \prod_{i=1,\dots,N; j=1,\dots,N'} (1 - \lambda_{i,p}\mu_{j,p}p^{-s})^{-1}.$$

The function $L^S(\boldsymbol{\tau} \times \boldsymbol{\tau}', s)$ is defined by meromorphic continuation on the whole complex plane. Moreover, it has no pole at $s = 1/2$. In the case $\boldsymbol{\tau} = \boldsymbol{\tau}'$, we can decompose this L-function as a product

$$L^S(\boldsymbol{\tau} \times \boldsymbol{\tau}, s) = L^S(\boldsymbol{\tau}, Sym^2, s)L^S(\boldsymbol{\tau}, \wedge^2, s).$$

The two functions are also eulerian products for $Re(s) \gg 0$ and, for $p \notin S$, we have

$$L_p(\boldsymbol{\tau}, Sym^2, s) = \prod_{1 \le i \le j \le N} (1 - \lambda_{i,p}\lambda_{j,p}p^{-s})^{-1},$$
$$L_p(\boldsymbol{\tau}, \wedge^2, s) = \prod_{1 \le i < j \le N} (1 - \lambda_{i,p}\lambda_{j,p}p^{-s})^{-1}.$$

The two functions have also meromorphic continuation to the whole complex plane. Put

$$L^S(\pi \times \sigma, s) = \prod_{i=1,\dots,k; j=1,\dots,l} L^S(\boldsymbol{\pi}_i \times \boldsymbol{\sigma}_j, s),$$
$$L^S(\pi, Ad, s) = \left(\prod_{1 \le i < i' \le k} L^S(\boldsymbol{\pi}_i \times \boldsymbol{\pi}_{i'}, s)\right)\left(\prod_{i=1,\dots,k} L^S(\boldsymbol{\pi}_i, \wedge^2, s)\right),$$
$$L^S(\sigma, Ad, s) = \left(\prod_{1 \le j < j' \le l} L^S(\boldsymbol{\sigma}_j \times \boldsymbol{\sigma}_{j'}, s)\right)\left(\prod_{j=1,\dots,l} L^S(\boldsymbol{\sigma}_i, Sym^2, s)\right).$$

In the forthcoming paper [A], Arthur will prove that the two last functions have no poles at $s = 1$ (and it is know they have no zero at this point). Put

$$P^S(\pi \times \sigma, s) = \frac{L^S(\pi \times \sigma, s)}{L^S(\pi, Ad, s + 1/2)L^S(\sigma, Ad, s + 1/2)}.$$

This function is holomorphic at $s = 1/2$.

For $v \in S$, we fix some invariant positive hermitian products on the spaces of E_{π_v} and E_{σ_v} and we define the semi-linear form $\mathcal{L}_{\pi_v,\sigma_v}$ as in Section 2.8 (in this section, v was finite, but the definition is the same if $v = \infty$).

We define the integral $J(\psi, \varphi)$ as in 1.1, and we denote by $||\varphi||$ and $||\psi||$ the L^2-norm of φ and ψ. For these definitions, we use the Tamagawa measures on $G(\mathbb{Q})\backslash G(\mathbb{A})$ and $H(\mathbb{Q})\backslash H(\mathbb{A})$. Remark that, for $v \in S$, a measure on $H(\mathbb{Q}_v)$ is used to define the form $\mathcal{L}_{\pi_v,\sigma_v}$. We assume that the Tamagawa measure is equal to the product of these measures for $v \in S$ and of the "unramified" measures for $v \notin S$ (these measures gives the measure 1 to an hyperspecial compact subgroup). Let

$$\Delta^S = \zeta^S(2)\dots\zeta^S(d-2)L^S(\delta, d/2),$$

where $\zeta^S(s)$ is the partial Riemann's ζ-function and $L^S(\delta, s)$ is the partial Dirichlet's function associated to the quadratic character determined by δ. At last, let

$$b = \begin{cases} k+l, & \text{if all } d_i \text{ are even,} \\ k+l-1, & \text{if at least one } d_i \text{ is odd.} \end{cases}$$

Then Ichino and Ikeda conjecture the following equality

$$\frac{|J(\psi,\varphi)|^2}{||\varphi||^2||\psi||^2} = 2^{-b}\Delta^S P^S(\pi\times\sigma, 1/2)\prod_{v\in S}\frac{\mathcal{L}_{\pi_v,\sigma_v}(e_v, e_v, e'_v, e'_v)}{(e_v, e_v)(e'_v, e'_v)}.$$

The right member does not depend on the choice of S.

3. An integral formula for ϵ-factors

3.1. The twisted group $GL(\mathbf{d})$. We come back to a local p-adic ground field F. We consider the group $GL(\mathbf{d})$ over F, where $\mathbf{d} \in \mathbb{N}$. Let $J_{\mathbf{d}}$ be the antidiagonal matrix

$$J_{\mathbf{d}} = \begin{pmatrix} & & & & 1 \\ & & & -1 & \\ & & 1 & & \\ & \cdot & & & \\ \cdot & & & & \\ (-1)^{\mathbf{d}-1} & & & & \end{pmatrix}.$$

Let $\theta_{\mathbf{d}}$ be the automorphism $g \mapsto J_{\mathbf{d}}{}^t g^{-1} J_{\mathbf{d}}^{-1}$ of $GL(\mathbf{d})$. We introduce the semi-direct product $GL(\mathbf{d})^+ = GL(\mathbf{d}) \rtimes \{1, \theta_{\mathbf{d}}\}$ and its connected component $GL(\mathbf{d})\theta_{\mathbf{d}}$. We call this component a twisted space over $GL(\mathbf{d})$. As we will see later, the theory of twisted endoscopy relates the harmonic analysis on classical groups, in particular on special orthogonal groups, with the harmonic analysis on such twisted spaces $GL(\mathbf{d})\theta_{\mathbf{d}}$. It is useful to describe $GL(\mathbf{d})\theta_{\mathbf{d}}$ in terms of elementary linear algebra. Fix a vector space $\mathbf{V}$ over F of dimension $\mathbf{d}$, with a basis $(v_i)_{i=1,\dots,\mathbf{d}}$. We identify as usual $GL(\mathbf{d})$ with the group $\mathbf{G}$ of automorphims of $\mathbf{V}$. Define a bilinear form $\boldsymbol{\theta}_{\mathbf{d}}$ on $\mathbf{V}$ by

$$\boldsymbol{\theta}_{\mathbf{d}}\left(\sum_{i=1,..,\mathbf{d}} x_i v_i, \sum_{i=1,\dots,\mathbf{d}} y_i v_i\right) = (-1)^{[(d+1)/2]}\sum_{i=1,\dots,\mathbf{d}}(-1)^i x_{\mathbf{d}+1-i}y_i.$$

Let's remark that $\boldsymbol{\theta}_{\mathbf{d}}$ is symmetric if $\mathbf{d}$ is odd and anti-symmetric if $\mathbf{d}$ is even. For $g \in GL(\mathbf{d})$, define the bilinear form $g\boldsymbol{\theta}_{\mathbf{d}}$ on $\mathbf{V}$ by

$$(g\boldsymbol{\theta}_{\mathbf{d}})(v, v') = \boldsymbol{\theta}_{\mathbf{d}}(g^{-1}v, v')$$

for every $v, v' \in \mathbf{V}$. The application $g\theta_{\mathbf{d}} \mapsto g\boldsymbol{\theta}_{\mathbf{d}}$ is a bijection between $GL(\mathbf{d})\theta_{\mathbf{d}}$ and the set $\tilde{\mathbf{G}}$ of non-degenerate bilinear forms on $\mathbf{V}$. The right and left actions of $GL(\mathbf{d})$ on $GL(\mathbf{d})\theta_{\mathbf{d}}$ become the actions of $\mathbf{G}$ on this set $\tilde{\mathbf{G}}$ of bilinear forms given by the formula

$$(g\gamma g')(v,v') = \gamma(g^{-1}v, g'v),$$

for every bilinear form γ, every $g, g' \in \mathbf{G}$ and every $v, v' \in \mathbf{V}$.

Each $\gamma \in \tilde{\mathbf{G}}$ defines an automorphism θ_γ of $\mathbf{G}$: for $g \in \mathbf{G}$, $\theta_\gamma(g)$ is the unique element of $\mathbf{G}$ such that $\theta_\gamma(g)\gamma = \gamma g$. Let $\mathbf{T}$ be a maximal torus of $\mathbf{G}$ defined over F and let $\mathbf{B}$ be a Borel subgroup containing $\mathbf{T}$ but not necessarily defined over F. Let $\tilde{\mathbf{T}}$ the set of $\gamma \in \tilde{\mathbf{G}}$ such that θ_γ preserves $\mathbf{T}$ and $\mathbf{B}$. It is a principal homogeneous space over $\mathbf{T}$ (for the left or right action), in particular $\mathbf{T}$ is well determined by $\tilde{\mathbf{T}}$. For $\gamma \in \tilde{\mathbf{T}}$, the restriction of θ_γ to $\mathbf{T}$ does not depend of γ. We denote it by $\theta_{\tilde{\mathbf{T}}}$. By definition, a maximal twisted subtorus of $\tilde{\mathbf{G}}$ is such set $\tilde{\mathbf{T}}$ satisfying the condition $\tilde{\mathbf{T}}(F) \neq \emptyset$. Let $A_{\tilde{\mathbf{T}}}$ the maximal subtorus split over F and contained in the subset of elements of $\mathbf{T}$ fixed by $\theta_{\tilde{\mathbf{T}}}$. We say that $\tilde{\mathbf{T}}$ is anisotropic if $A_{\tilde{\mathbf{T}}} = \{1\}$.

3.2. Embedding of twisted spaces. Let $\mathbf{d}$ and $\mathbf{r}$ be two natural integers such that $2\mathbf{r}+1 \leq \mathbf{d}$. Let $\mathbf{V}$ be a vector space over F of dimension $\mathbf{d}$ and let $\mathbf{V} = \mathbf{W} \oplus \mathbf{Z}$ be a direct decomposition, where $dim(\mathbf{Z}) = 2\mathbf{r}+1$. Denote $\mathbf{G}$, resp. $\mathbf{H}$, the group of linear automorphisms of $\mathbf{V}$, resp. $\mathbf{W}$, and denote $\tilde{\mathbf{G}}$, resp. $\tilde{\mathbf{H}}$, the set of non-degenerate bilinear forms on $\mathbf{V}$, resp. $\mathbf{W}$. There is an embedding of $\mathbf{H}$ in $\mathbf{G}$: we extend every automorphism of $\mathbf{W}$ as an automorphism of $\mathbf{V}$ that acts by identity on $\mathbf{Z}$. Fix $\boldsymbol{\nu} \in F^\times$ and a quadratic form $\zeta_{\mathbf{Z}}$ on $\mathbf{Z}$ which is a direct sum of hyperbolic spaces and of the one dimensional form

$$(x,y) \mapsto \boldsymbol{\nu} xy.$$

Then there is an embedding of $\tilde{\mathbf{H}}$ in $\tilde{\mathbf{G}}$: we extend each bilinear form on $\mathbf{W}$ as the direct sum of this form on $\mathbf{W}$ and the quadratic form $\zeta_{\mathbf{Z}}$ on $\mathbf{Z}$. The two embeddings are compatible with the actions of $\mathbf{G}$ and $\mathbf{H}$ on $\tilde{\mathbf{G}}$ and $\tilde{\mathbf{H}}$.

3.3. The integral formula for twisted spaces. Let $\boldsymbol{\pi}$ be a smooth irreducible representation of $GL_{\mathbf{d}}(F)$. Assume that $\boldsymbol{\pi} \circ \theta_{\mathbf{d}} \simeq \boldsymbol{\pi}$ (we say $\boldsymbol{\pi}$ is autodual). Then we can extend $\boldsymbol{\pi}$ in a representation $\boldsymbol{\pi}^+$ of $GL^+_{\mathbf{d}}(F)$. There is two possible extensions and we must choose one. For that, assume that $\boldsymbol{\pi}$ is generic. That means that there exists a non-zero linear form ϕ on the space $E_{\boldsymbol{\pi}}$ of $\boldsymbol{\pi}$ such that

$$\phi(\boldsymbol{\pi}(u)e) = \psi_F\left(\sum_{i=1,\dots,\mathbf{d}-1} u_{i,i+1}\right)\phi(e)$$

for every $e \in E_{\boldsymbol{\pi}}$ and every unipotent matrix

$$u = \begin{pmatrix} 1 & u_{12} & u_{13} & . & . \\ 0 & 1 & u_{23} & . & . \\ . & . & . & . & . \\ . & . & . & . & . \\ . & . & 0 & 0 & 1 \end{pmatrix}.$$

Such ϕ is called a Whittaker's functional. As it is well know, the space of Whittaker's functionals is a line. For each of the two extensions $\boldsymbol{\pi}^+$, the operator $\boldsymbol{\pi}^+(\theta_{\mathbf{d}})$ preserves this line and acts on it by a sign ± 1. We choose $\boldsymbol{\pi}^+$ such that the sign is $+1$. Let $\tilde{\boldsymbol{\pi}}$ be the restriction of $\boldsymbol{\pi}^+$ to $GL_{\mathbf{d}}(F)\theta_{\mathbf{d}}$.

Similarly, let $\boldsymbol{\sigma}$ be a smooth irreducible representation of $GL_{\mathbf{d}'}(F)$, where $\mathbf{d}' = \mathbf{d} - 2\mathbf{r} - 1$. We assume that $\boldsymbol{\sigma} \circ \theta_{\mathbf{d}'} \simeq \boldsymbol{\sigma}$ and that $\boldsymbol{\sigma}$ is generic. Then we define a representation $\boldsymbol{\sigma}^+$ of $\mathbf{G}^+_{\mathbf{d}'}(F)$ and its restriction $\tilde{\boldsymbol{\sigma}}$ to $GL_{\mathbf{d}'}(F)\theta_{\mathbf{d}'}$.

By a choice of basis of $\mathbf{V}$ and $\mathbf{W}$, we identify $GL_{\mathbf{d}}(F)\theta_{\mathbf{d}}$ with $\tilde{\mathbf{G}}(F)$ and $GL_{\mathbf{d}'}(F)\theta_{\mathbf{d}'}$ with $\tilde{\mathbf{H}}(F)$.

Consider data $(\mathbf{W}', \mathbf{W}'', \tilde{\mathbf{T}}', \zeta_{\mathbf{W}''})$ where:

- $\mathbf{W}'$ and $\mathbf{W}''$ are supplementary subspaces of $\mathbf{W}$;
- denote by $\mathbf{H}'$ the group of automorphisms of $\mathbf{W}'$ and by $\tilde{\mathbf{H}}'$ the set of non-degenerate bilinear forms on $\mathbf{W}'$; then $\tilde{\mathbf{T}}'$ is a maximal twisted subtorus of $\tilde{\mathbf{H}}'$;
- $\zeta_{\mathbf{W}''}$ is a non-degenerate quadratic form on $\mathbf{W}''$.

We impose to those data the following conditions:

(1) the dimension of $\mathbf{W}'$ is even;
(2) the torus $\tilde{\mathbf{T}}'$ is anisotropic;
(3) the special orthogonal group $SO(\zeta_{\mathbf{W}''})$ of the quadratic form $\zeta_{\mathbf{W}''}$ is quasi-split over F and the special orthogonal $SO(\zeta_{\mathbf{V}''})$ of the quadratic form $\zeta_{\mathbf{V}''} = \zeta_{\mathbf{W}''} \oplus \zeta_{\mathbf{Z}}$ on $\mathbf{V}'' = \mathbf{W}'' \oplus \mathbf{Z}$ is also quasi-split.

Under those assumptions, we denote by $\tilde{\mathbf{T}}$ the subset of $\tilde{\mathbf{H}}$ whose elements are direct sums of $\zeta_{\mathbf{W}''}$ on $\mathbf{W}''$ and a bilinear form on $\mathbf{W}'$ which belongs to $\tilde{\mathbf{T}}'$. This set is like a non-maximal twisted subtorus of $\tilde{\mathbf{H}}$.

Denote by $\mathbf{T}$ the maximal subtorus of $\mathbf{H}'$ associated to $\tilde{\mathbf{T}}'$. The group $\mathbf{T}(F)$ acts by conjugacy on $\tilde{\mathbf{T}}(F)$. We denote by $\tilde{\mathbf{T}}(F)_{/\theta}$ the set of conjugacy classes. Denote by $\mathbf{T}_\theta$ the identity component of the subset of elements of $\mathbf{T}$ fixed by $\theta_{\tilde{\mathbf{T}}'}$. Because of (2), it is an anisotropic torus and we normalize an Haar measure on $\mathbf{T}_\theta(F)$ by $mes(\mathbf{T}_\theta(F)) = 1$. Let $\gamma \in \tilde{\mathbf{T}}(F)_{/\theta}$. The application

$$\begin{array}{ccc} \mathbf{T}_\theta(F) & \to & \tilde{\mathbf{T}}(F)_{/\theta} \\ t & \mapsto & t\gamma \end{array}$$

is a local isomorphism near $1 \in \mathbf{T}_\theta(F)$. We define the mesure on $\tilde{\mathbf{T}}(F)_{/\theta}$ such that it is invariant by translations by $\mathbf{T}(F)$ and that the application above preserves the measures near $1 \in \mathbf{T}_\theta(F)$.

The normalizer $Norm_{\mathbf{H}}(\tilde{\mathbf{T}})$ of $\tilde{\mathbf{T}}$ in $\mathbf{H}$ contains $\mathbf{T} \times SO(\zeta_{\mathbf{W}''})$. Let

$$W(H, \tilde{\mathbf{T}}) = Norm_{\mathbf{H}}(\tilde{\mathbf{T}})(F)/(\mathbf{T}(F) \times SO(\zeta_{\mathbf{W}''})(F)).$$

It is a finite group.

Now we define some functions almost everywhere on $\tilde{\mathbf{T}}(F)$. Let $\gamma \in \tilde{\mathbf{T}}(F)$ in general position. The connected centralizer of γ in $\mathbf{G}$ is the group $\mathbf{T}_\theta \times SO(\zeta_{\mathbf{V}''})$. Thanks to Clozel, the representations of non-connected groups have the same properties as in the connected case. In particular, they have characters and their characters have near a semi-simple point a development as in 2.3. Then we can define $c_{\tilde{\boldsymbol{\pi}}}(\gamma)$ as we have defined $c_{\pi}(t)$ in 2.4. We define similarly $c_{\tilde{\boldsymbol{\sigma}}}(\gamma)$. Let ${}^t\gamma$ the bilinear form deduced by symmetry from γ: ${}^t\gamma(v,v') = \gamma(v',v)$. Restrict the two forms γ and ${}^t\gamma$ to $\mathbf{W}'$. Consider the quotient

$$|det(\gamma_{|\mathbf{W}'} - {}^t\gamma_{|\mathbf{W}'})|_F |det(\gamma_{|\mathbf{W}'})|_F^{-1},$$

where the determinants are computed respectively to some fixed basis of $\mathbf{W}'$. It does not depend of the basis and is non-zero for γ in general position. We let

$$\Delta_{\mathbf{r}}(\gamma) = |2|_F^{\mathbf{r}^2+\mathbf{r}+\mathbf{r}dim(\mathbf{W}'')} |det(\gamma_{|\mathbf{W}'} - {}^t\gamma_{|\mathbf{W}'})|_F^{\mathbf{r}} |det(\gamma_{|\mathbf{W}'})|_F^{-\mathbf{r}}.$$

At last, let

$$D^{\tilde{\mathbf{H}}}(\gamma) = |det((1-\theta_\gamma)_{|\mathfrak{h}/\mathfrak{h}_\gamma})|_F,$$

where $\mathfrak{h}$ is the Lie algebra of $\mathbf{H}$ and $\mathfrak{h}_\gamma$ is the Lie algebra of the centralizer of γ in $\mathbf{H}$.

Consider the set of subsets $\tilde{\mathbf{T}}$ obtained by the procedure above. The group $\mathbf{H}(F)$ acts by conjugacy on this set. Let $\boldsymbol{\mathcal{T}}$ a set of representatives of conjugacy classes. Now we can define

$$\epsilon_{geom}(\tilde{\boldsymbol{\pi}}, \tilde{\boldsymbol{\sigma}}) = \sum_{\tilde{\mathbf{T}} \in \boldsymbol{\mathcal{T}}} |W(\mathbf{H}, \tilde{\mathbf{T}})|^{-1} \int_{\tilde{\mathbf{T}}(F)_{/\theta}} c_{\tilde{\boldsymbol{\pi}}}(\gamma) c_{\tilde{\boldsymbol{\sigma}}}(\gamma) D^{\tilde{\mathbf{H}}}(\gamma) \Delta_{\mathbf{r}}(\gamma)\, d\gamma.$$

3.4. Computation of some value of ϵ-factors by an integral formula. Recall that Jacquet, Piatetski-Shapiro and Shalika have defined an L-function $L(\boldsymbol{\pi} \times \boldsymbol{\sigma}, s)$ and an ϵ-factor $\epsilon(\boldsymbol{\pi} \times \boldsymbol{\sigma}, s, \psi_F)$, where s is a complex parameter ([JPSS] theorem 2.7). Recall also that $\boldsymbol{\pi}$ and $\boldsymbol{\sigma}$ have central characters $\omega_{\boldsymbol{\pi}}$ and $\omega_{\boldsymbol{\sigma}}$. They are quadratic characters because we have assumed $\boldsymbol{\pi}$ and $\boldsymbol{\sigma}$ autodual.

PROPOSITION. *Assume $\boldsymbol{\pi}$ and $\boldsymbol{\sigma}$ autodual and tempered. Then there is an equality*

$$\omega_{\boldsymbol{\pi}}(\boldsymbol{\nu}(-1)^{[\mathbf{d}'/2]}) \omega_{\boldsymbol{\sigma}}(\boldsymbol{\nu}(-1)^{1+[\mathbf{d}/2]}) \epsilon(\boldsymbol{\pi} \times \boldsymbol{\sigma}, 1/2, \psi_F) = \epsilon_{geom}(\tilde{\boldsymbol{\pi}}, \tilde{\boldsymbol{\sigma}}).$$

Cf. [W3] théorème 7.1.

REMARKS. (1) Because $\boldsymbol{\pi}$ and $\boldsymbol{\sigma}$ are tempered, they are also generic and we can normalize $\tilde{\boldsymbol{\pi}}$ and $\tilde{\boldsymbol{\sigma}}$ as in 3.3.

(2) The terms $\boldsymbol{\nu}$ and ψ_F appear explicitely in the left member. But the right member depends also of $\boldsymbol{\nu}$ and ψ_F: $\tilde{\boldsymbol{\pi}}$ and $\tilde{\boldsymbol{\sigma}}$ are normalized using Whittaker's functionals, and this notion depends on ψ_F; and $\boldsymbol{\nu}$ is used to define the embedding $\tilde{\mathbf{H}} \to \tilde{\mathbf{G}}$.

(3) At the present time, there is a gap in the proof. We use some results coming from the twisted local trace formula and this formula is not yet

written in the litterature. We have some reasons to think that it will be proved soon.

The proof follows the same lines as that of proposition 2.5. I want explain only why appears the ϵ-factor. I assume here $\mathbf{d}' = \mathbf{d} - 1$. As for special orthogonal groups, the space $Hom_{\mathbf{H}(F)}(E_{\boldsymbol{\pi}}, E_{\boldsymbol{\sigma}})$ plays a crucial role. Aizenbud, Gourevitch, Rallis and Shiffmann have proved that its dimension is not greater as 1. We see below that it is equal to 1. Assuming this result, let l be a non-zero element of this space and let γ be an element of $\tilde{\mathbf{H}}(F)$. The linear form $\tilde{\boldsymbol{\sigma}}(\gamma)^{-1} \circ l \circ \tilde{\boldsymbol{\pi}}(\gamma)$ does not depend of γ. In fact, if we replace γ by $\gamma' = h\gamma$, with $h \in \mathbf{H}(F)$, we have

$$\begin{aligned}\tilde{\boldsymbol{\sigma}}(\gamma')^{-1} \circ l \circ \tilde{\boldsymbol{\pi}}(\gamma') &= \tilde{\boldsymbol{\sigma}}(\gamma)^{-1} \circ \boldsymbol{\sigma}(h)^{-1} \circ l \circ \boldsymbol{\pi}(h) \circ \tilde{\boldsymbol{\pi}}(\gamma) \\ &= \tilde{\boldsymbol{\sigma}}(\gamma)^{-1} \circ l \circ \tilde{\boldsymbol{\pi}}(\gamma),\end{aligned}$$

because $\boldsymbol{\sigma}(h)^{-1} \circ l \circ \boldsymbol{\pi}(h) = l$. It is easy to see that $\tilde{\boldsymbol{\sigma}}(\gamma)^{-1} \circ l \circ \tilde{\boldsymbol{\pi}}(\gamma)$ is also an element of $Hom_{\mathbf{H}(F)}(E_{\boldsymbol{\pi}}, E_{\boldsymbol{\sigma}})$. So there exists $c \in \mathbb{C}^{\times}$, independant of γ, such that

$$\tilde{\boldsymbol{\sigma}}(\gamma)^{-1} \circ l \circ \tilde{\boldsymbol{\pi}}(\gamma) = cl.$$

A computation similar as that of 2.9 holds, where this constant c plays the same role as the multiplicity $m(\rho, \sigma)$ before. But, as we see below, the constant c is equal to $\epsilon(\boldsymbol{\pi} \times \boldsymbol{\sigma}, 1/2, \psi_F)$, up to elementary terms. To see that, we must recall some results of Jacquet, Piatetski-Shapiro and Shalika. At this point, it is more clear to denote $\iota : \mathbf{H} \to \mathbf{G}$ and $\tilde{\iota} : \tilde{\mathbf{H}} \to \tilde{\mathbf{G}}$ the two embeddings of 3.2. Fix a basis $(v_i)_{i=1,\ldots,\mathbf{d}}$ of $\mathbf{V}$ such that $(v_i)_{i=1,\ldots,\mathbf{d}-1}$ is a basis of $\mathbf{W}$ and $v_{\mathbf{d}}$ is a basis of $\mathbf{Z}$. Using those basis, we identify $\mathbf{G}$ with $GL(\mathbf{d})$ and $\mathbf{H}$ with $GL(\mathbf{d} - 1)$. The embedding ι is

$$\iota : h \mapsto \begin{pmatrix} h & 0 \\ 0 & 1 \end{pmatrix}.$$

Fix a Whittaker's functional $\phi \neq 0$ on $E_{\boldsymbol{\pi}}$. For $e \in E_{\boldsymbol{\pi}}$, we define a function W_e on $GL(\mathbf{d}, F)$ by

$$W_e(g) = \phi(\boldsymbol{\pi}(g)e).$$

Similarly, we fix a non-zero Whittaker's functional on $E_{\boldsymbol{\sigma}}$ and, for $e' \in E_{\boldsymbol{\sigma}}$, we define a function $W_{e'}$ on $GL(\mathbf{d} - 1, F)$. For $s \in \mathbb{C}$, let

$$\mathcal{L}(W_{e'}, W_e, s) = \int_{GL(\mathbf{d}-1,F)} \overline{W_{e'}(h)} W_e(\iota(h)) |det(h)|_F^{s-1/2} dh.$$

Jacquet, Piatetski-Shapiro and Shalika have proved the three following properties ([JPSS], theorem 2.7):

(4) this integral is absolutely convergent for $Re(s) > 0$ and can be extended as a meromorphic function of s on the whole complex plane;

(5) we can choose e and e' such that $\mathcal{L}(W_{e'}, W_e, 1/2) \neq 0$;

(6) let $w_{\mathbf{d}}$ be the antidiagonal matrix of rank $\mathbf{d}$ with coefficients 1 on the antidiagonal and define

$$\check{W}_e(g) = W_e(w_{\mathbf{d}}{}^t g^{-1});$$

define similarly the function $\check{W}_{e'}$; then we have the functional equation

$$\begin{aligned}&\mathcal{L}(\check{W}_{e'}, \check{W}_e, 1-s)/L(1-s, \boldsymbol{\pi} \times \boldsymbol{\sigma})\\ &\quad = \omega_{\boldsymbol{\sigma}}(-1)^{\mathbf{d}-1}\epsilon(\boldsymbol{\pi} \times \boldsymbol{\sigma}, s, \psi_F)\mathcal{L}(W_{e'}, W_e, s)/L(s, \boldsymbol{\pi} \times \boldsymbol{\sigma})\end{aligned}$$

(here, we use the fact that $\boldsymbol{\pi}$ and $\boldsymbol{\sigma}$ are unitary and autodual).

Now, fix an invariant definite-positive hermitian form $(.,.)$ on $E_{\boldsymbol{\sigma}}$. We can define a linear application $l : E_{\boldsymbol{\pi}} \to E_{\boldsymbol{\sigma}}$ by

$$(e', l(e)) = \mathcal{L}(W_{e'}, W_e, 1/2).$$

We see that $l \in Hom_{\mathbf{H}(F)}(E_{\boldsymbol{\pi}}, E_{\boldsymbol{\sigma}})$. Then the property (5) shows that this space is non-zero. To compute the constant c, we can choose $\gamma = \boldsymbol{\theta}_{\mathbf{d}-1}$, so that

$$\tilde{\boldsymbol{\sigma}}(\boldsymbol{\theta}_{\mathbf{d}-1})^{-1} \circ l \circ \tilde{\boldsymbol{\pi}}(\tilde{\iota}(\boldsymbol{\theta}_{\mathbf{d}-1})) = cl.$$

This relation is equivalent to

$$(e_1', l(e_1)) = c(e', l(e)),$$

where

$$e_1' = \tilde{\boldsymbol{\sigma}}(\boldsymbol{\theta}_{\mathbf{d}-1})e'$$

and

$$e_1 = \tilde{\boldsymbol{\pi}}(\tilde{\iota}(\boldsymbol{\theta}_{\mathbf{d}-1}))e,$$

or to

$$\mathcal{L}(W_{e_1'}, W_{e_1}, 1/2) = c\mathcal{L}(W_{e'}, W_e, 1/2). \tag{7}$$

A tedious computation shows that there exists two matrices $a, b \in GL(\mathbf{d} - 1, F)$ such that, for every $h \in GL(\mathbf{d} - 1, F)$,

$$W_{e_1'}(h) = \check{W}_{e'}(z'ahb), \ W_{e_1}(\iota(h)) = \check{W}_e(z\iota(ahb)),$$

where z' and z are explicit central matrices of ranks $\mathbf{d} - 1$ and $\mathbf{d}$. The formulas above are equivalent to

$$W_{e_1'}(h) = \omega_{\boldsymbol{\sigma}}(z')\check{W}_{e'}(ahb), \ W_{e_1}(\iota(h)) = \omega_{\boldsymbol{\pi}}(z)\check{W}_e(\iota(ahb)).$$

By the change of variables $h \mapsto a^{-1}hb^{-1}$, we obtain that

$$\mathcal{L}(W_{e_1'}, W_{e_1}, 1/2) = \overline{\omega_{\boldsymbol{\sigma}}(z')}\omega_{\boldsymbol{\pi}}(z)\mathcal{L}(\check{W}_{e'}, \check{W}_e, 1/2).$$

Then the relations (6) and (7) imply

$$c = \overline{\omega_{\boldsymbol{\sigma}}(z')}\omega_{\boldsymbol{\pi}}(z)\omega_{\boldsymbol{\sigma}}(-1)^{\mathbf{d}-1}\epsilon(\boldsymbol{\pi} \times \boldsymbol{\sigma}, 1/2, \psi_F).$$

That is the expected relation between c and the ϵ-factor.

4. Endoscopy and twisted endoscopy

4.1. Remark about notations. In the Gross-Prasad conjecture, there is two special orthogonal groups of two spaces of distinct parities. Sometimes, it is better to use a notation that makes clear what are the bigger group and the bigger space and what are the smaller group and the smaller space. In this case, we use the notation of Section 2: G is the bigger group, V is the bigger space, H is the smaller group and W is the smaller space. But sometimes, it is better to use a notation that makes clear what is the even-dimensional space and what is the odd-dimensional space. In this case, we use the following notation: V is the even-dimensional space, d is its dimension and G is its special orthogonal group; V' is the odd-dimensional space, d' is its dimension and G' is its special orthogonal group. In the present section, we use this notation.

4.2. Endoscopy in odd dimensional case. Consider a vector space V' over F, of finite odd dimensional dimension d', with a non-degenerate quadratic form q', and denote by G' its special orthogonal group. Let $g \in G'(F)$ be a strongly regular element (that means its centralizer is a maximal subtorus of G'). The eigenvalues in $\bar{F}$ of g (acting on $V' \otimes_F \bar{F}$) are all distinct. The set $\Lambda(g)$ of those eigenvalues contains 1 and is invariant by $\lambda \mapsto \lambda^{-1}$. We can write

$$\Lambda(g) = \{1, \lambda_1^{\pm 1}, \ldots, \lambda_{(d'-1)/2}^{\pm 1}\}.$$

Define the stable conjugacy class $C^{st}(g)$ of g as the set of $g' \in G'(F)$ for which there exists $x \in G'(\bar{F})$ such that $g' = x^{-1}gx$. Then, for $g' \in G'(F)$, g' belongs to $C^{st}(g)$ if and only if g' is strongly regular and $\Lambda(g') = \Lambda(g)$. It is easy to prove that the set of ordinary conjugacy classes by $G'(F)$ contained in $C^{st}(g)$ is finite.

Consider now two others vector spaces V_1' and V_2' over F, of finite odd dimension d_1' and d_2', with non-degenerate quadratic forms q_1' and q_2', and denote by G_1' and G_2' their special orthogonal groups. We assume

(1) $d_1' + d_2' = d' + 1$;

(2) G_1' and G_2' are split over F.

Let $g \in G'(F)$, $g_1 \in G_1'(F)$ and $g_2 \in G_2'(F)$ three strongly regular elements. We say that g and (g_1, g_2) correspond each other, or that $C^{st}(g)$ and $C^{st}(g_1) \times C^{st}(g_2)$ correspond each other, if and only if $\Lambda(g) = \Lambda(g_1) \cup \Lambda(g_2)$ (then the intersection of the two last sets is reduced to $\{1\}$). It is clear that, given g_1 and g_2, there is at most one stable conjugacy class $C^{st}(g)$ that corresponds to $C^{st}(g_1) \times C^{st}(g_2)$. In the opposite direction, given g, there is at most a finite set of stable conjugacy classes $C^{st}(g_1) \times C^{st}(g_2)$ that correspond to $C^{st}(g)$.

Let g, g_1 and g_2 be as above, assume that g corresponds to (g_1, g_2). Then Langlands and Shelstad have defined a transfer factor $\Delta_{G_1' \times G_2', G'}(g_1, g_2; g) \in \mathbb{C}^\times$. The general definition gives such factor only up to homothety, but in our particular case, there is a natural normalization which gives a well defined transfer factor. It is a sign $\Delta_{G'_1\times G'_2,G'}(g_1,g_2;g)\in\{\pm 1\}$. As function of (g_1,g_2), it depends only of the stable conjugacy classes $C^{st}(g_1)\times C^{st}(g_2)$. It is more subtle as function of g. It depends only of the ordinary conjugacy class by $G'(F)$ of g. The set of ordinary conjugacy classes contained in $C^{st}(g)$ has a natural structure of principal homogeneous space over some cohomology group, which is isomorphic to a finite product of copies of $\mathbb{Z}/2\mathbb{Z}$. The datum (g_1,g_2) determines some character of this group and the function $g\mapsto\Delta_{G'_1\times G'_2,G'}(g_1,g_2;g)$ transforms according to this character.

Now consider a function θ on $G'(F)$ which is a finite linear combination of characters of irreducible smooth representations of $G'(F)$. Consider similar functions θ_1 on $G'_1(F)$ and θ_2 on $G'_2(F)$. Such functions are invariant by ordinary conjugacy. Assume that θ_1 and θ_2 are stable, that means they are constant on every stable conjugacy class. We say that θ is a transfert of $\theta_1\times\theta_2$ if and only if, for every strongly regular $g\in G'(F)$, we have the equality

$$(3)\qquad D^{G'}(g)^{1/2}\theta(g)=\sum_{(g_1,g_2)}D^{G'_1}(g_1)^{1/2}D^{G'_2}(g_2)^{1/2}\times\Delta_{G'_1\times G'_2,G'}(g_1,g_2;g)\theta_1(g_1)\theta_2(g_2),$$

where the sum is over the pairs (g_1,g_2) up to stable conjugacy such that $C^{st}(g)$ and $C^{st}(g_1)\times C^{st}(g_2)$ correspond each other, and the functions $D^{G'}$, $D^{G'_1}$ and $D^{G'_2}$ are defined as in 2.4. For instance

$$D^{G'}(g)=|det((ad(g)-1)_{|\mathfrak{g}'(F)/\mathfrak{g}'_g(F)})|_F.$$

REMARK. We can reverse the order of G'_1 and G'_2. If g corresponds to $(g_1,g_2)\in G'_1(F)\times G'_2(F)$, it is clear that g corresponds also to $(g_2,g_1)\in G'_2(F)\times G'_1(F)$. But $\Delta_{G'_1\times G'_2,G'}(g_1,g_2;g)$ is not always equal to $\Delta_{G'_2\times G'_1,G'}(g_2,g_1;g)$. In fact, we have

$$(4)\qquad \Delta_{G'_2\times G'_1,G'}(g_2,g_1;g)=\mu(G')\Delta_{G'_1\times G'_2,G'}(g_1,g_2;g),$$

where

$$\mu(G')=\begin{cases}1, & \text{if } G' \text{ is split;}\\ -1, & \text{if } G' \text{ is not split.}\end{cases}$$

In certain sense, we can consider that $G'_1\times G'_2$ and $G'_2\times G'_1$ are equivalent, but for us, it is better to distinguish the two pairs.

4.3. Twisted endoscopy in the odd dimensional case. Let V', q' and G' be as in the preceding section. Here, we assume that G' is split. There are two useful cases of twisted endoscopy. Let's begin with the most usual. Let $\mathbf{V}$ be a vector space over F of even dimension $\mathbf{d}=d'-1$. As in 3.1, we define the group $\mathbf{G}=GL(\mathbf{V})$ and the twisted space $\tilde{\mathbf{G}}$ of non-degenerate bilinear forms on $\mathbf{V}$. For $\gamma\in\tilde{\mathbf{G}}(F)$, let $x_\gamma:\mathbf{V}\to\mathbf{V}^*$ and $x^*_\gamma:\mathbf{V}\to\mathbf{V}^*$ be the two dual isomorphisms such that

$$\gamma(v,v')=<v,x_\gamma(v')>=<x^*_\gamma(v),v'>$$

for every $v, v' \in \mathbf{V}$, where $< .,. >$ is the natural pairing between $\mathbf{V}$ and $\mathbf{V}^*$. Define an automorphism y_γ of $\mathbf{V}$ by $y_\gamma = (x^*_\gamma)^{-1}x_\gamma$. We say that γ is strongly regular if and only if the centralizer of $\mathbf{G}_\gamma$ in $\mathbf{G}$ is a maximal subtorus of $\mathbf{G}$. It is equivalent to say that y_γ is strongly regular (its centralizer in $\mathbf{G}$ is a maximal subtorus). In this case, let $\Lambda(\gamma)$ be the set of eigenvalues in $\bar{F}$ of y_γ. It is invariant by $\lambda \mapsto \lambda^{-1}$. Define the stable conjugacy class of γ as the set of $\gamma' \in \tilde{\mathbf{G}}(F)$ for which there exists $x \in \mathbf{G}(\bar{F})$ such that $\gamma' = x^{-1}\gamma x$. An element $\gamma' \in \tilde{\mathbf{G}}(F)$ belongs to $C^{st}(\gamma)$ if and only if γ' is strongly regular and $\Lambda(\gamma') = \Lambda(\gamma)$.

Let $\gamma \in \tilde{\mathbf{G}}(F)$ and $g \in G'(F)$ two strongly regular elements. We say that γ and g correspond each other, or that $C^{st}(\gamma)$ and $C^{st}(g)$ correspond each other if and only if they satisfy the equality

$$\Lambda(\gamma) = \{-\lambda; \lambda \in \Lambda(g), \lambda \neq 1\}.$$

The change of signs is related to the fact that the basic element $\boldsymbol{\theta}_{\mathbf{d}}$ is symplectic. This correspondence is a bijection between stable conjugacy classes of strongly regular elements in $\tilde{\mathbf{G}}(F)$ and in $G'(F)$.

Let γ and g as before, assume they correspond each other. Then Kottwitz and Shelstad have defined a transfer factor which is a sign $\Delta_{G',\tilde{\mathbf{G}}}(g,\gamma) \in \{\pm 1\}$. In fact, in our situation, it is always 1. Consider a function θ on $G'(F)$ which is a finite linear combination of characters of smooth irreducible representations of $G'(F)$ (we apologize for the double meaning of the symbol θ). Assume θ is stable. As in 3.1, we can identify $\tilde{\mathbf{G}}$ with a component of the non-connected group $GL(\mathbf{d})^+$. Consider a function $\tilde{\boldsymbol{\theta}}$ on $\tilde{\mathbf{G}}(F)$ which is the restriction to $\tilde{\mathbf{G}}(F)$ of a finite linear combination of characters of smooth irreducible representations of $GL(\mathbf{d})^+(F)$. We say that $\tilde{\boldsymbol{\theta}}$ is a transfer of θ if and only if for every strongly regular $\gamma \in \tilde{\mathbf{G}}(F)$, we have the equality

$$(1) \qquad |2|_F^a D^{\tilde{\mathbf{G}}}(\gamma)^{1/2}\tilde{\boldsymbol{\theta}}(\gamma) = \sum_g D^{G'}(g)^{1/2}\Delta_{G',\tilde{\mathbf{G}}}(g;\gamma)\theta(g),$$

where the sum is over the strongly regular elements $g \in G'(F)$, up to stable conjugacy, that correspond to γ, where the functions $D^{\tilde{\mathbf{G}}}$ and $D^{G'}$ are as in 3.3 and 2.4 and where $a = -(d-1)/4$. Because the correspondence between stable conjugacy classes is a bijection, the sum in g is in fact reduced to a unique term.

Remark. Let $\mathbf{T}$ be the centralizer of $\mathbf{G}_\gamma$ in $\mathbf{G}$. Then we have

$$|2|_F^a = |det((1-\theta_\gamma)_{|\mathbf{t}/\mathbf{g}_\gamma})|_F^{-1}.$$

Now consider the second useful case of twisted endoscopy. Let $\mathbf{V}'$ be a vector space over F of odd dimension $\mathbf{d}' = d'$. We introduce similarly the group $\mathbf{G}' = GL(\mathbf{V}')$ and the twisted space $\tilde{\mathbf{G}}'$. For $\gamma \in \tilde{\mathbf{G}}'(F)$ we define as before the set $\Lambda(\gamma)$ and, if γ is strongly regular, the stable conjugacy class $C^{st}(\gamma)$. In this case, the set $\Lambda(\gamma)$ contains always 1. Let $\gamma \in \tilde{\mathbf{G}}'(F)$ and

$g \in G'(F)$ two strongly regular elements. We say that γ and g correspond each other, or that $C^{st}(\gamma)$ and $C^{st}(g)$ correspond each other if and only if they satisfy the equality $\Lambda(g) = \Lambda(\gamma)$. There is no sign here because $\boldsymbol{\theta}_{\mathbf{d}'}$ is now symmetric.

Let γ and g as before, assume they correspond each other. Then Kottwitz and Shelstad have defined a transfer factor which is a sign $\Delta_{G',\tilde{\mathbf{G}}'}(g,\gamma) \in \{\pm 1\}$. It is more complicated as in the preceding case. The factor $\Delta_{G',\tilde{\mathbf{G}}'}(g;\gamma)$ depends only of the stable conjugacy class of g and of the ordinary conjugacy class by $\mathbf{G}'(F)$ of γ. As in the preceding section, the set of ordinary conjugacy classes contained in $C^{st}(\gamma)$ has a natural structure of principal homogeneous space over some cohomology group, which is isomorphic to a finite product of copies of $\mathbb{Z}/2\mathbb{Z}$. The datum g determines some character of this group and the function $\gamma \mapsto \Delta_{G',\tilde{\mathbf{G}}'}(g;\gamma)$ transforms according to this character.

Consider a function θ on $G'(F)$ and a function $\tilde{\boldsymbol{\theta}}$ on $\tilde{\mathbf{G}}'(F)$ satisfying similar properties as above. We say that $\tilde{\boldsymbol{\theta}}$ is a transfer of θ if and only if for every strongly regular $\gamma \in \tilde{\mathbf{G}}(F)$, we have the equality (1), where now $a = -(d+1)/4$ (the sum in g is again reduced to a unique term).

4.4. Endoscopy in the even dimensional case. Consider a vector space V over F, of finite even dimension d, provided with a non-degenerate quadratic form q, and let G be its special orthogonal group. Let $g \in G(F)$ be a strongly regular element. The eigenvalues in $\bar{F}$ of g are all distinct. The set $\Lambda(g)$ of those eigenvalues is of the form

$$\Lambda(g) = \{\lambda_1^{\pm 1}, \dots, \lambda_{d/2}^{\pm 1}\}.$$

Define the stable conjugacy class $C^{st}(g)$ of g as in 4.2 . Then, if g' belongs to $C^{st}(g)$, we have $\Lambda(g') = \Lambda(g)$. But this condition is not sufficient. In fact the set of strongly regular $g' \in G(F)$ such that $\Lambda(g') = \Lambda(g)$ is the union of two stable conjugacy classes, which are conjugate by an element of determinant -1 of the full orthogonal group.

Consider two others vector spaces V_1 and V_2 over F, of finite even dimension d_1 and d_2, with non-degenerate quadratic forms q_1 and q_2, and denote by G_1 and G_2 their special orthogonal groups. We assume

(1) $d_1 + d_2 = d$;
(2) G_1 and G_2 are quasi-split over F;
(3) $\delta = \delta_1\delta_2$, where $\delta, \delta_1, \delta_2 \in F^{\times}/F^{\times,2}$ are the discriminants of q, q_1 and q_2 (by convention, $\delta_i = 1$ if $d_i = 0$ for $i = 1$ or 2).

Let $g \in G(F)$, $g_1 \in G_1(F)$ and $g_2 \in G_2(F)$ three strongly regular elements. In first approximation, we can say that g and (g_1, g_2) correspond each other, or that $C^{st}(g)$ and $C^{st}(g_1) \times C^{st}(g_2)$ correspond each other, if and only if $\Lambda(g) = \Lambda(g_1) \cup \Lambda(g_2)$.But this correspondence is not sufficiently fine, because the sets of eigenvalues does not determine the stable conjugacy classes. We must define a finer correspondence. I will give very briefly the definition.

We fix once for all maximal subtori T of G, T_1 of G_1, T_2 of G_2, defined over $\bar{F}$. Consider the group $X^*(T)$ of algebraic characters of T, defined over $\bar{F}$. We can fix a basis $\mathcal{B}$ of $X^*(T)$ over $\mathbb{Z}$ such that, for every $t \in T(\bar{F})$, the set of eigenvalues of the action of t in $V \otimes_F \bar{F}$ is

$$\{(x^*(t))^{\pm 1}; x^* \in \mathcal{B}\}.$$

We fix similarly basis $\mathcal{B}_1$ of $X^*(T_1)$ and $\mathcal{B}_2$ of $X^*(T_2)$. Next, fix a bijection between $\mathcal{B}$ and $\mathcal{B}_1 \sqcup \mathcal{B}_2$. By linearity, it defined an isomorphism $X^*(T) \to X^*(T_1) \oplus X^*(T_2)$, then an isomorphism $\iota : T_1 \times T_2 \to T$, defined over $\bar{F}$. Denote by W the Weyl group of G relative to T. Our element g is conjugate in $G(\bar{F})$ to an element $t \in T(\bar{F})$. Similarly, g_1 and g_2 are conjugate to $t_1 \in T_1(\bar{F})$ and $t_2 \in T_2(\bar{F})$. Then we say that g and (g_1, g_2) correspond each other if and only if the images of t and of $\iota(t_1, t_2)$ in T/W are equal.

This correspondence has similar properties as in the odd dimensional case.

As we said in 4.2, the general definition of transfer factor gives only a function up to homothety. To normalize it, some choice are needed. I don't explain that, but it appears that the choice of an element $\nu \in F^\times/F^{\times,2}$ is sufficient to normalize the transfer factor. Then, we choose such ν and we have a transfer factor $\Delta_{G_1 \times G_2, G}(g_1, g_2; g) \in \{\pm 1\}$, g, g_1 and g_2 being as above, assuming that g corresponds to (g_1, g_2).

It is convenient to define a sign $\mu(G)$ similar to the sign defined in 4.2. The definition depends now on ν. Consider the line $Z(\nu) = F$ with the quadratic form $q_{Z(\nu)}(x, x_1) = -2\nu x x_1$. Denote by $G'(\nu)$ the special orthogonal group of the orthogonal sum $V \oplus Z(\nu)$. We define

$$\mu(G) = \begin{cases} 1, & \text{if } G'(\nu) \text{ is split over } F; \\ -1, & \text{if } G'(\nu) \text{ is not split over } F. \end{cases}$$

Recall that the elementary theory of quadratic forms over local fields says that V is always the orthogonal direct sum of hyperbolic planes and of an anisotropic space of dimension $d_{an} = 0$, 2 or 4. When $d_{an} = 0$, then G is split and $G'(\nu)$ is also split whatever ν. When $d_{an} = 4$, then G is not quasi-split and $G'(\nu)$ is not split whatever ν. But if $d_{an} = 2$, then G is quasi-split and $G'(\nu)$ can be split or not split according to ν.

Now consider a function θ on $G(F)$ which is a finite linear combination of characters of irreducible smooth representations of $G(F)$. Consider similar functions θ_1 on $G_1(F)$ and θ_2 on $G_2(F)$. Assume that θ_1 and θ_2 are stable. We say that θ is a transfert of $\theta_1 \times \theta_2$ if and only if, for every strongly regular $g \in G(F)$, we have an equality similar to 4.2 (3).

Remark. As in the odd-dimensional case, we can reverse the order of G_1 and G_2. An equality similar to 4.2 (4) remains true.

4.5. Twisted endoscopy in the even-dimensional case. Let V, q and G be as in the preceding section. Here, we assume that G is quasi-split. Let $\mathbf{V}$ be a vector space over F of dimension $\mathbf{d} = d$. We introduce the group $\mathbf{G} = GL(\mathbf{V})$ and the twisted space $\tilde{\mathbf{G}}$. For a strongly regular element $\gamma \in \tilde{\mathbf{G}}(F)$, we define the set $\Lambda(\gamma)$ as in 4.3 .

Let $\gamma \in \tilde{\mathbf{G}}(F)$ and $g \in G(F)$ two strongly regular elements. We say that γ and g corresponds each other, or that $C^{st}(\gamma)$ and $C^{st}(g)$ corresponds each other if and only if they satisfy the equality

$$\Lambda(\gamma) = \{-\lambda; \lambda \in \Lambda(g)\}.$$

In fact, for each $C^{st}(g)$, there is a unique corresponding $C^{st}(\gamma)$. The application $C^{st}(g) \mapsto C^{st}(\gamma)$ is not surjective. On its image, the fibers have two elements (the two stable conjugacy classes determined by the set of eigenvalues).

Let γ and g as before, assume they correspond each other. Then Kottwitz and Shelstad have defined a transfer factor $\Delta_{G,\tilde{\mathbf{G}}}(g, \gamma) \in \{\pm 1\}$ (that is not always 1). Consider a function θ on $G(F)$ which is a finite linear combination of characters of smooth irreducible representations of $G(F)$. Assume θ is stable. Consider a function $\tilde{\boldsymbol{\theta}}$ on $\tilde{\mathbf{G}}(F)$ which is, as in 4.3, the restriction to $\tilde{\mathbf{G}}(F)$ of a finite linear combination of characters of smooth irreducible representations of $GL(\mathbf{d})^+(F)$. We say that $\tilde{\boldsymbol{\theta}}$ is a transfer of θ if and only if for every strongly regular $\gamma \in \tilde{\mathbf{G}}(F)$, an equality similar to 4.3 (1) is satisfied, where $a = -d/4$.

4.6. Parametrization and endoscopy: the odd-dimensional case. We can now state more a precise version of the conjectures of 1.3. Let V', q', G' be as in 4.2 (d' is odd). We use the definitions and notations of 1.3. Let $\varphi' \in \Phi^{orth}_{temp}(d')$. The central elements ± 1 of $Sp(d'-1, \mathbb{C})$ are contained in $S(\varphi')$. Denote by $z_{\varphi'}$ the image of -1 in $S(\varphi')/S(\varphi')^0$. For $\mu \in \{\pm 1\}$, denote by $\mathcal{E}^{\mu}(\varphi')$ the set of characters ϵ' of $S(\varphi')/S(\varphi')^0$ such that $\epsilon'(z_{\varphi'}) = \mu$. We have stated that to φ' is associated a packet $\Pi^{G'}(\varphi')$ of irreducible tempered representations of $G'(F)$. We add now that there is a bijection

$$(1) \qquad \begin{aligned} \mathcal{E}^{\mu(G')}(\varphi') &\to \Pi^{G'}(\varphi') \\ \epsilon' \quad &\mapsto \pi(\varphi', \epsilon'). \end{aligned}$$

Let's remark that $\mathcal{E}^{-}(\varphi')$ can be empty (when the center of $Sp(d'-1, \mathbb{C})$ in included in $S(\varphi')^0$). In this case, $\Pi^{G'}(\varphi')$ is empty if $\mu(G') = -1$.

For $s' \in S(\varphi')/S(\varphi')^0$, we define the following linear combination of characters of irreducible representations

$$(2) \qquad \theta^{G'}(\varphi', s') = \sum_{\epsilon' \in \mathcal{E}^{\mu(G')}(\varphi')} \epsilon'(s')\theta_{\pi(\varphi', \epsilon')}.$$

Assume $\mu(G') = 1$. Then G' is split. In this case, a first required property is

(3) the functions $\theta^{G'}(\varphi', 1_{\varphi'})$ is stable, where $1_{\varphi'}$ is the identity element of $S(\varphi')/S(\varphi')^0$.

Denote by $\boldsymbol{\varphi}' : W_{DF} \to GL(d'-1, \mathbb{C})$ the composition of φ' and of the natural embedding $Sp(d'-1, \mathbb{C}) \to GL(d'-1, \mathbb{C})$. The local conjecture of Langlands for the group $GL(d'-1)$ has been proved by Harris-Taylor and Henniart ([HT], [H]). It associates to $\boldsymbol{\varphi}'$ an irreducible representation $\boldsymbol{\pi}(\boldsymbol{\varphi}')$ of $GL(d'-1, F)$. This representation is tempered and autodual, so we can extend it as in 3.3 in a representation $\boldsymbol{\pi}(\boldsymbol{\varphi}')^+$ of $GL(d'-1)^+(F)$. We denote by $\tilde{\boldsymbol{\pi}}(\boldsymbol{\varphi}')$ its restriction to $GL(d'-1, F)\theta_{d-1}$. Introducing a space $\mathbf{V}$ of dimension $\mathbf{d} = d'-1$ as in 4.3, we identify $\theta_{\tilde{\boldsymbol{\pi}}(\boldsymbol{\varphi}')}$ to a function on $\tilde{\mathbf{G}}(F)$. We require

(4) there is some scalar $c(\varphi') \in \mathbb{C}$ such that $|c(\varphi')| = 1$ and that $c(\varphi')\theta_{\tilde{\boldsymbol{\pi}}(\boldsymbol{\varphi}')}$ is the transfer of $\theta^{G'}(\varphi', 1_{\varphi'})$.

Denote by $\boldsymbol{\varphi}'_> : W_{DF} \to GL(d', \mathbb{C})$ the direct sum of $\boldsymbol{\varphi}'$ and of the trivial one-dimensional representation of W_{DF}. We define similarly the function $\theta_{\tilde{\boldsymbol{\pi}}(\boldsymbol{\varphi}'_>)}$ on $\tilde{\mathbf{G}}'(F)$, where $\tilde{\mathbf{G}}'$ is as in 4.3 (then $\mathbf{G}' \simeq GL(d')$). We require

(5) there is some scalar $c_>(\varphi') \in \mathbb{C}$ such that $|c_>(\varphi')| = 1$ and that $c_>(\varphi')\theta_{\tilde{\boldsymbol{\pi}}(\boldsymbol{\varphi}'_>)}$ is the transfer of $\theta^{G'}(\varphi', 1_{\varphi'})$.

We have stated the assertion (4) because it is the most usual form of twisted endoscopy for odd-dimensional special orthogonal groups. But for our problem, we use only the assertion (5).

Now we remove the assumption $\mu(G') = 1$. Consider two groups G'_1 and G'_2 as in 4.2. Let $\varphi'_1 \in \Phi^{orth}_{temp}(d'_1)$ and $\varphi'_2 \in \Phi^{orth}_{temp}(d'_2)$. There is a natural embedding

$$Sp(d'_1 - 1, \mathbb{C}) \times Sp(d'_2 - 1, \mathbb{C}) \to Sp(d' - 1, \mathbb{C})$$

well defined up to conjugacy. Assume that φ' is equal to the composition of (φ'_1, φ'_2) with this embedding. Consider the image in $Sp(d'-1, \mathbb{C})$ of the product of the central element $1 \in Sp(d'_1 - 1, \mathbb{C})$ and of the central element $-1 \in Sp(d'_2 - 1, \mathbb{C})$. It belongs to $S(\varphi')$. Denote by s' its image in $S(\varphi')/S(\varphi')^0$. Then we require

(6) for every $\mu \in \{\pm 1\}$, there is some scalar $\gamma^\mu(\varphi'_1, \varphi'_2) \in \mathbb{C}$ such that $|\gamma^\mu(\varphi'_1, \varphi'_2)| = 1$ and that $\gamma^{\mu(G')}(\varphi'_1, \varphi'_2)\theta^{G'}(\varphi', s')$ is the transfer of $\theta^{G'_1}(\varphi'_1, 1_{\varphi'_1}) \times \theta^{G'_2}(\varphi'_2, 1_{\varphi'_2})$.

Let's remark that the scalars $c(\varphi')$ and $c_>(\varphi')$ of (4) and (5) are well determined. But the assertion (6) is less precise: we can change the scalar $\gamma^\mu(\varphi'_1, \varphi'_2)$ if we change the bijections (1). In the opposite direction, if we fix the scalars $\gamma^\mu(\varphi', \varphi'')$ for all possible μ, G'_1, G'_2, φ'_1, φ'_2, the bijections (1) is well determined. In fact, assuming the preceding assertions, the following proposition fix the scalars.

PROPOSITION. *The scalars $c_>(\varphi')$ of (5) are equal to* 1. *Up to change of the bijections (1), we can assume that the scalars $\gamma^\mu(\varphi'_1, \varphi'_2)$ of (6) are equal to μ.*

It is convenient to state here this proposition, but we prove it at the same time as the main theorem 5.4 below.

4.7. Parametrization and endoscopy: the even-dimensional case. Let V, q, G be as in 4.4 . The dimension d is even. We must consider the complex group $SO(d,\mathbb{C})$ as the Langlands dual of G. This statement contains some implicit data. In practice, those data are as follows. We fix a maximal subtorus T of G defined over $\bar{F}$ and a basis $\mathcal{B}$ of $X^*(T)$ satisfying the same properties as in 4.4 . The group $SO(d,\mathbb{C})$ is a subgroup of $GL(d,\mathbb{C})$ and acts on $\mathbb{C}^d$. Fix a maximal subtorus $\hat{T}$ of $SO(d,\mathbb{C})$ and a basis $\hat{\mathcal{B}}$ of $X^*(\hat{T})$ such that, for every $\hat{t}\in\hat{T}$, the set of eigenvalues of the action of $\hat{t}$ in $\mathbb{C}^d$ is

$$\{(x^*(\hat{t}))^{\pm 1}; x^*\in\hat{\mathcal{B}}\}.$$

The essential data is a perfect duality between $X^*(T)$ and $X^*(\hat{T})$ for which $\mathcal{B}$ and $\hat{\mathcal{B}}$ are dual basis.

Recall that $\delta\in F^\times/F^{\times,2}$ is the discriminant of q. Let $\varphi\in\Phi^{orth}_{temp}(d,\delta)$. The central elements ± 1 of $SO(d,\mathbb{C})$ are contained in $S(\varphi)$. Denote by z_φ the image of -1 in $S(\varphi)/S(\varphi)^0$. For $\mu\in\{\pm 1\}$, denote by $\mathcal{E}^\mu(\varphi)$ the set of characters ϵ of $S(\varphi)/S(\varphi)^0$ such that $\epsilon(z_\varphi)=\mu$. To φ is associated a packet $\Pi^G(\varphi)$ of irreducible tempered representations of $G(F)$. We require that there exists a bijection

$$(1)\qquad \begin{array}{ccc}\mathcal{E}^{\mu(G)}(\varphi) & \to & \Pi^G(\varphi)\\ \epsilon & \mapsto & \pi(\varphi,\epsilon).\end{array}$$

Remark. We have said in 4.4 that, if $d_{an}(V)=0$, resp. 4, $\mu(G)=1$, resp. -1. In this case, the set $\mathcal{E}^{\mu(G)}(\varphi)$ is well defined. But, if $d_{an}(V)=2$, $\mu(G)$ can be ± 1 according to the choice of ν. Then, according to this choice, the same set $\Pi^G(\varphi)$ can be parametrized by the two distinct sets $\mathcal{E}^+(\varphi)$ or $\mathcal{E}^-(\varphi)$.

For $s\in S(\varphi)/S(\varphi)^0$, we define $\theta^G(\varphi,s)$ by an equality similar to 4.6 (2).
Assume first $\mu(G)=1$. Then G is quasi-split. As in 4.6, we require
(2) the functions $\theta^G(\varphi,1_\varphi)$ is stable.

Denote by $\boldsymbol{\varphi}: W_{DF}\to GL(d,\mathbb{C})$ the composition of φ with of the natural embedding $O(d,\mathbb{C})\to GL(d,\mathbb{C})$. As in 4.6, we define a function $\theta_{\tilde{\pi}(\boldsymbol{\varphi})}$ on $\tilde{\mathbf{G}}(F)$, where $\tilde{\mathbf{G}}$ is as in 4.5. We require

(3) there is some scalar $c(\varphi)\in\mathbb{C}$ such that $|c(\varphi)|=1$ and that $c(\varphi)\theta_{\tilde{\pi}(\boldsymbol{\varphi})}$ is the transfer of $\theta^G(\varphi,1_\varphi)$.

Now we remove the assumption $\mu(G)=1$. Consider two groups G_1 and G_2 as in 4.4. Let $\varphi_1\in\Phi^{orth}_{temp}(d_1,\delta_1)$ and $\varphi_2\in\Phi^{orth}_{temp}(d_2,\delta_2)$. There is a natural embedding

$$(4)\qquad O(d_1,\mathbb{C})\times O(d_2,\mathbb{C})\to O(d,\mathbb{C}),$$

but well defined only up to conjugacy by the full orthogonal group $O(d,\mathbb{C})$. We need to be more precise. We have fixed above some data: T, $\mathcal{B}$, $\hat{T}$, $\hat{\mathcal{B}}$ and a perfect duality between $X^*(T)$ and $X^*(\hat{T})$. Of course, we assume fixed similar data T_1, $\mathcal{B}_1$ etc... for the other groups. In 4.4, we have fixed data T, $\mathcal{B}$, T_1 etc... We assume there are the same as before. Then, using the three perfect dualities, the isomorphism $\iota : T_1 \times T_2 \to T$ of 4.4 defines an isomorphism $\hat{\iota} : \hat{T} \to \hat{T}_1 \times \hat{T}_2$. We choose the embedding (4) such that its restriction to $\hat{T}_1 \times \hat{T}_2$ is $\hat{\iota}^{-1}$.

Assume that φ is equal to the composition of (φ_1, φ_2) and this embedding. Consider the image in $SO(d, \mathbb{C})$ of the product of the central element $1 \in SO(d_1, \mathbb{C})$ and of the central element $-1 \in SO(d_2, \mathbb{C})$. It belongs to $S(\varphi)$. Denote by s its image in $S(\varphi)/S(\varphi)^0$. Then we require

(5) for every $\mu \in \{\pm 1\}$, there is some scalar $\gamma^\mu(\varphi_1, \varphi_2) \in \mathbb{C}$ such that $|\gamma^\mu(\varphi_1, \varphi_2)| = 1$ and that $\gamma^{\mu(G)}(\varphi_1, \varphi_2)\theta^G(\varphi, s)$ is the transfer of $\theta^{G_1}(\varphi_1, 1_{\varphi_1}) \times \theta^{G_2}(\varphi_2, 1_{\varphi_2})$.

As in 4.6, we can be more precise concerning the scalars. Recall that the center of $G(F)$ is $\{\pm 1\}$ if $d \neq 2$ and, in any case, contains $\{\pm 1\}$. Then, for every irreducible smooth representation π of $G(F)$, the value $\omega_\pi(-1)$ of the central character of π at the central element -1 is well defined, and, of course, it is a sign ± 1.

PROPOSITION. *(i) For every $\varphi \in \Phi^{orth}_{temp}(d, \delta)$, there is a sign $\zeta(\varphi) = \pm 1$ such that for every $\pi \in \Pi^G(\varphi)$, we have $\omega_\pi(-1) = \zeta(\varphi)$ (independently of $\mu(G)$).*

(ii) The scalar $c(\varphi)$ of (3) is equal to $\zeta(\varphi)\epsilon(\boldsymbol{\pi}(\boldsymbol{\varphi}), 1/2, \psi_F)^{-1}$.

(iii) Up to change of the bijections (1), we can assume that the scalars $\gamma^\mu(\varphi_1, \varphi_2)$ of (5) are all equal to 1.

The first assertion is proved directly. The two others are proved at the same time as the main theorem.

REMARK. (6) As in 3.4, it is reasonable that the scalar $c(\varphi)$ depends on ψ_F because the normalization of $\tilde{\boldsymbol{\pi}}(\boldsymbol{\varphi})$ depends on it.

(7) $\epsilon(\boldsymbol{\pi}(\boldsymbol{\varphi}), 1/2, \psi_F)$ is always a four root of unity, but not always ± 1. So the same holds for the scalar $c(\varphi)$.

4.8. The present state of the conjectures. Arthur has now proved a substantial part of the conjectures stated in the Sections 4.2 to 4.7. The proofs will appear in the forthcoming paper [A]. More precisely, in the odd-dimensional case, he has proved all conjectures for split groups G'. There is no doubt that the same methods can be used in the non-split case, and may be the final Arthur's article will include this case. In the even-dimensional case, there is a slight problem: it is difficult to distinguish two representations of $G(F)$ that are not equivalent but are conjugate by an element of the full orthogonal group. Similarly, it is difficult to distinguish two parameters φ that are not equivalent modulo conjugacy by $SO(d, \mathbb{C})$, but becomes equivalent modulo conjugacy by $O(d, \mathbb{C})$. It is possible to state weaker conjectures where we identify two representations of $G(F)$ conjugate by an element of the full orthogonal group and, similarly, we identify two parameters conjugate by an element of $O(d,\mathbb{C})$. Arthur prove this weaker conjecture when the group G is quasi-split. As in the odd-dimensional case, the same proof holds certainly if G is not quasi-split. I don't know if it is possible to prove the stronger form of the conjectures without substantial additional work.

5. The precise form of the Gross-Prasad conjecture

5.1. Description of some centralizers. Let $N \geq 1$ be an integer and let $\varphi : W_{DF} \to GL(N,\mathbb{C})$ be a continuous semi-simple homomorphism whose restriction to $SL(2,\mathbb{C})$ is algebraic. Assume that either the image of φ in contained in $O(N,\mathbb{C})$, or N is even and the image of φ is contained in $Sp(N,\mathbb{C})$. Then φ is conjugate to the homomorphism $w \mapsto \varphi^\theta(w) = {}^t\varphi(w)^{-1}$. Consider an irreducible component ψ of φ, which takes its values in some group $GL(N_\psi,\mathbb{C})$. Then ψ^θ is also such irreducible component. If $\psi^\theta \simeq \psi$, its implies that, up to conjugacy, either ψ takes its values in $O(N_\psi,\mathbb{C})$, or N_ψ is even and ψ takes its values in $Sp(N_\psi,\mathbb{C})$ (the two cases are disjoint). Then we can decompose φ as

$$\varphi = (\oplus_{i\in I^{orth}} l_i\varphi_i) \oplus (\oplus_{i\in I^{symp}} l_i\varphi_i) \oplus (\oplus_{j\in J} l_j(\varphi_j \oplus \varphi_j^\theta)),$$

where:

- I^{orth}, I^{symp} and J are finite disjoint sets;
- for $i \in I^{orth}$, φ_i is an irreducible homomorphism from W_{DF} to some $O(N_i,\mathbb{C})$ and $l_i \geq 1$ is its multiplicity; if $i \neq i'$, φ_i is not equivalent to $\varphi_{i'}$;
- for $i \in I^{symp}$, φ_i is an irreducible homomorphism from W_{DF} to some $Sp(N_i,\mathbb{C})$ with N_i even, and $l_i \geq 1$ is its multiplicity; if $i \neq i'$, φ_i is not equivalent to $\varphi_{i'}$;
- for $j \in J$, φ_j is an irreducible homomorphism from W_{DF} to some $GL(N_j,\mathbb{C})$ and φ_j^θ is not equivalent to φ_j; $l_j \geq 1$ is its multiplicity; if $j \neq j'$, φ_j is equivalent neither to $\varphi_{j'}$ nor to $\varphi_{j'}^\theta$.

It is easy to compute the centralizer in $GL(N,\mathbb{C})$ of the image of φ. It is isomorphic to

$$\left(\prod_{i\in I^{orth}} GL(l_i,\mathbb{C})\right) \times \left(\prod_{i\in I^{symp}} GL(l_i,\mathbb{C})\right) \times \left(\prod_{j\in J}(GL(l_j,\mathbb{C}) \times GL(l_j,\mathbb{C}))\right).$$

Consider more precisely the case where N is even and φ takes its values in $Sp(N,\mathbb{C})$. The centralizer in $Sp(N,\mathbb{C})$ of the image of φ is the intersection with $Sp(N,\mathbb{C})$ of the group above. Let $i \in I^{orth}$. The group $GL(l_i,\mathbb{C})$ that appears is in fact the group of automorphisms of the space U_i of intertwining homomorphisms from φ_i to φ. Because the space of φ_i is provided with a quadratic form and the one of φ is provided with a symplectic form, there is a natural symplectic form on U_i. This imply first that l_i is even. Then the intersection of $GL(l_i,\mathbb{C})$ with $Sp(N,\mathbb{C})$ is the symplectic group of this symplectic form, in other words a group $Sp(l_i,\mathbb{C})$. Now let $i \in I^{symp}$. For the same reason, the space U_i defined as before is provided with a quadratic form, and the intersection of $GL(l_i,\mathbb{C})$ with $Sp(N,\mathbb{C})$ is isomorphic to the orthogonal group of this quadratic form, in other words to a group $O(l_i,\mathbb{C})$. Lastly, let $j \in J$. Then the group $GL(l_j,\mathbb{C}) \times GL(l_j,\mathbb{C})$ that appears is the product of the groups of automorphisms of two spaces U_j and U_j^θ, where U_j, resp. U_j^θ, is the space of intertwining homomorphisms from φ_j, resp. φ_j^θ, to φ. The symplectic form of the space of φ induces a duality between U_j and U_j^θ. The intersection of $GL(l_j,\mathbb{C}) \times GL(l_j,\mathbb{C})$ with $Sp(N,\mathbb{C})$ is the subgroup of elements compatible with this duality. Its group is isomorphic with $GL(l_i,\mathbb{C})$. We obtain that the centralizer in $Sp(N,\mathbb{C})$ of the image of φ is

$$\left(\prod_{i\in I^{orth}} Sp(l_i,\mathbb{C})\right) \times \left(\prod_{i\in I^{symp}} O(l_i,\mathbb{C})\right) \times \left(\prod_{j\in J} GL(l_j,\mathbb{C})\right).$$

Its group of connected components is therefore isomorphic to $\{\pm 1\}^{I^{symp}}$.

Assume now that φ takes its values in $O(N,\mathbb{C})$. A similar computation holds for the centralizer in $O(N,\mathbb{C})$ of the image of φ. We obtain

$$\left(\prod_{i\in I^{orth}} O(l_i,\mathbb{C})\right) \times \left(\prod_{i\in I^{symp}} Sp(l_i,\mathbb{C})\right) \times \left(\prod_{j\in J} GL(l_j,\mathbb{C})\right).$$

But we are more interested by the centralizer in $SO(N,\mathbb{C})$ and not in the full orthogonal group. A little computation shows that it is the subgroup of elements x such that, if $(x_i)_{i\in I^{orth}}$ are the components of x in the first product, we have $\prod_{i\in I^{orth}} det(x_i)^{N_i} = 1$. The group of connected components of this centralizer is therefore isomorphic to the subgroup of the elements $s = (s_i)_{i\in I^{orth}} \in \{\pm 1\}^{I_{orth}}$ such that $\prod_{i\in I^{orth}, N_i \text{ odd}} s_i = 1$.

5.2. Definition of some sign. Let N and N' be two natural integers, with N' even. Let $\varphi : W_{DF} \to GL(N,\mathbb{C})$ and $\varphi' : W_{DF} \to GL(N',\mathbb{C})$ be two homomorphisms as in the preceding section. Assume the image of φ is contained in $O(N,\mathbb{C})$ and the image of φ' is contained in $Sp(N',\mathbb{C})$. By the Langlands correspondence, we associate to these homomorphisms the representations $\pi(\varphi)$ of $GL(N,F)$ and $\pi(\varphi')$ of $GL(N',F)$. As in 1.3, we denote by δ_φ the quadratic character of W_{DF} defined by $\delta_\varphi = det \circ \varphi$. By Kummer's theory, we can see δ_φ as a quadratic character of $F^\times$. Define

$$E(\varphi,\varphi') = \delta_\varphi(-1)^{N'/2}\epsilon(\pi(\varphi) \times \pi(\varphi'), 1/2, \psi_F).$$

LEMMA. *The number $E(\varphi,\varphi')$ belongs to $\{\pm 1\}$ and does not depend of ψ_F.*

5.3. Definition of some character. Let d and d' be two natural numbers, with d even and d' odd. Let $\delta \in F^\times/F^{\times,2}$, $\varphi \in \Phi^{orth}(d,\delta)$ and $\varphi' \in \Phi^{orth}(d')$. As in the preceding sections, we denote by $\boldsymbol{\varphi}$ the composition of φ with the embedding $O(d,\mathbb{C}) \to GL(d,\mathbb{C})$ and by $\boldsymbol{\varphi}'$ the composition of φ' with the embedding $Sp(d'-1,\mathbb{C}) \to GL(d'-1,\mathbb{C})$. We apply the constructions of 5.1 , denoting by I^{orth}, I^{symp} etc... the data relative to $\boldsymbol{\varphi}$ and by I'^{orth}, I'^{symp} etc... those relative to $\boldsymbol{\varphi}'$. Then we obtain

$$S(\varphi')/S(\varphi')^0 = \{\pm 1\}^{I'^{symp}}$$

and

$$S(\varphi)/S(\varphi)^0 \subset \{\pm 1\}^{I^{orth}}.$$

Denote by $< .,. >$ the duality between $\{\pm 1\}$ and itself, that is

$$<\eta,\eta'> = \begin{cases} -1, & \text{if } \eta = \eta' = -1, \\ 1, & \text{otherwise.} \end{cases}$$

Define a character $\underline{\epsilon}'$ of $S(\varphi')/S(\varphi')^0 = \{\pm 1\}^{I'^{symp}}$ by

$$\underline{\epsilon}'(s') = \prod_{i' \in I'^{symp}} < s'_{i'}, E(\boldsymbol{\varphi}, \boldsymbol{\varphi}'_{i'}) >$$

for every $s' = (s'_{i'})_{i' \in I'^{symp}} \in \{\pm 1\}^{I'^{symp}}$. Define a character $\underline{\epsilon}$ of $\{\pm 1\}^{I^{orth}}$ by

$$\underline{\epsilon}(s) = \prod_{i \in I^{orth}} < s_i, E(\boldsymbol{\varphi}_i, \boldsymbol{\varphi}') >$$

for every $s = (s_i)_{i \in I^{orth}} \in \{\pm 1\}^{I^{orth}}$. By restriction to $S(\varphi)/S(\varphi)^0$, we obtain a character of this group, also denoted by $\underline{\epsilon}$.

We have defined in 4.6 and 4.7 two elements $z_\varphi \in S(\varphi)/S(\varphi)^0$ and $z_{\varphi'} \in S(\varphi')/S(\varphi')^0$. A little computation shows that

$$(1) \qquad \underline{\epsilon}(z_\varphi) = \underline{\epsilon}'(z_{\varphi'}) = E(\boldsymbol{\varphi}, \boldsymbol{\varphi}').$$

5.4. Statement of the local Gross-Prasad conjecture. Let's come back to the situation of 2.1, but changing the notation as explained in 4.1. So the data $d = dim(V)$, G, $dim(W)$, H of 2.1 become d, G, d', G', if the bigger space is even-dimensional, and become d', G', d, G if the bigger space is odd-dimensional. For smooth irreducible representations π of $G(F)$ and π' of $G'(F)$, we denote by $m(\pi,\pi')$ the multiplicity defined in 2.1 if the first case and the multiplicity denoted in 2.1 by $m(\pi',\pi)$ in 2.1 in the other case.

Some element $\nu \in F^\times/F^{\times,2}$ occurs in 2.1 when we write explicitely the quadratic form of the space Z. In 4.4, we use some element ν to define $\mu(G)$. And this sign is used to parametrize the tempered representations of $G(F)$. We assume now that the two elements ν of 2.1 and 4.4 are the same. The space $V \oplus Z(\nu)$ of 4.4 is the same as V', up to addition of hyperbolic planes. Then $G'(\nu)$ is split if and only if G' is split. The definitions imply the equality

$$\mu(G) = \mu(G').$$

Let $\varphi \in \Phi^{orth}_{temp}(d,\delta)$ and let $\varphi' \in \Phi^{orth}_{temp}(d')$, where as before δ is the discriminant of the quadratic form on the even-dimensional space. To those parameters are associated packets $\Pi^G(\varphi)$ and $\Pi^{G'}(\varphi')$ of tempered representations of $G(F)$ and $G'(F)$. Those packets are themselves parametrized by the sets of characters $\mathcal{E}^{\mu(G)}(\varphi)$ and $\mathcal{E}^{\mu(G')}(\varphi')$.

REMARK. We assume the parametrizations fixed such that the equalities of propositions 4.6 and 4.7 are satisfied. There is no logical problem here: these propositions and the theorem below follow from the same proof, but the propositions are proved firstly.

Applying the contructions of 5.3, we define two characters $\underline{\epsilon}$ and $\underline{\epsilon}'$. The equality 5.3(1) and the equality above show that there are two possibilities:

(1) if $E(\boldsymbol{\varphi},\boldsymbol{\varphi}') = \mu(G) = \mu(G')$, then $\underline{\epsilon}$ belongs to $\mathcal{E}^{\mu(G)}(\varphi)$ and $\underline{\epsilon}'$ belongs to $\mathcal{E}^{\mu(G')}(\varphi')$;

(2) if $E(\boldsymbol{\varphi},\boldsymbol{\varphi}') = -\mu(G) = -\mu(G')$, then $\underline{\epsilon}$ does not belong to $\mathcal{E}^{\mu(G)}(\varphi)$ and $\underline{\epsilon}'$ does not belong to $\mathcal{E}^{\mu(G')}(\varphi')$.

THEOREM. *(i) If $E(\boldsymbol{\varphi},\boldsymbol{\varphi}') = \mu(G) = \mu(G')$, then $m(\pi(\varphi,\underline{\epsilon}),\pi(\varphi',\underline{\epsilon}')) = 1$ and $m(\pi,\pi') = 0$ for every $(\pi,\pi') \in \Pi^G(\varphi)\times\Pi^{G'}(\varphi')$ such that $(\pi,\pi') \neq (\pi(\varphi,\underline{\epsilon}),\pi(\varphi',\underline{\epsilon}'))$.*

(ii) If $E(\boldsymbol{\varphi},\boldsymbol{\varphi}') = -\mu(G) = -\mu(G')$, then $m(\pi,\pi') = 0$ for every $(\pi,\pi') \in \Pi^G(\varphi)\times\Pi^{G'}(\varphi')$.

Cf. [W4] théorème 4.9. Let's recall the theorem is conditional: we assume true the conjectures of Section 4, and in addition some results coming from the twisted local trace formula.

The theorem proves the local Gross-Prasad conjecture. Of course, it implies the theorem A and B of 1.5 and 1.6. There is two pairs (G_i,H_i) and (G_a,H_a) of groups in these theorems, and not one as in the statement above. We explain that in the remark 2 of the following section.

5.5. An idea of the proof. To state the preceding theorem, we started from given groups G and G'. In order to prove it, it is better to start from more abstract data. Those data are:

- an element $\nu \in F^\times/F^{\times,2}$ fixed once and for all;
- two natural integers d and d', with d even and d' odd;
- an element $\delta \in F^\times/F^{\times,2}$, with $\delta = 1$ if $d = 0$;
- a sign $\mu \in \{\pm 1\}$.

Let $\varphi \in \Phi^{orth}_{temp}(d,\delta)$, $\varphi' \in \Phi^{orth}_{temp}(d')$, $\epsilon \in \mathcal{E}^{\mu}(\varphi)$ and $\epsilon' \in \mathcal{E}^{\mu}(\varphi')$. We can define the multiplicity $m(\varphi,\epsilon;\varphi',\epsilon')$ as follows. Using the elementary classification's theory of quadratic forms over local fields, we see that, up to isometry, there exists exactly two spaces provided with non-degenerate quadratic forms satisfying the conditions of 2.1, where ν is our choosen ν, such that the even-dimensional space has dimension d and discriminant δ, the odd-dimensional space has dimension d' and the special orthogonal group G' of the odd-dimensional space is split if $\mu = 1$ and non-split if $\mu = -1$. We apply the constructions of the preceding sections to those spaces. The pairs (φ,ϵ) and (φ',ϵ') defines representations $\pi(\varphi,\epsilon)$ of $G(F)$ and $\pi(\varphi',\epsilon')$ of $G'(F)$ and we let

$$m(\varphi,\epsilon;\varphi',\epsilon') = m(\pi(\varphi,\epsilon),\pi(\varphi',\epsilon')).$$

REMARKS. (1) For defining the representation $\pi(\varphi,\epsilon)$ of $G(F)$, we need to consider $SO(d,\mathbb{C})$ as the Langlands dual group of G and for that reason, we need to fix some additional data, cf. 4.7. To change those data can replace $\pi(\varphi,\epsilon)$ by a representation conjugate by an element of the full orthogonal group. But fortunately, we see easily that the multiplicity above does not change. So the additional data are not necessary for our problem.

(2) To obtain the situation of theorems A and B, we fix only ν, d, d' and δ. Choosing $\mu = 1$, the preceding construction gives two groups that we denote by G_i and G'_i (or G_i and H_i with the notation of Section 1). Choosing $\mu = -1$, the construction gives other groups G_a and G'_a. The group G_i is quasi-split and the group G'_i is split. The group G_a is an inner form of G_i (that can be also quasi-split and in this case isomorphic to G_i) and the group G'_a is a non-quasi-split inner form of G'_i.

The problem is to compute $m(\varphi,\epsilon;\varphi',\epsilon')$ in terms of ϵ-factors. Let φ and φ' be as before and let $s \in S(\varphi)/S(\varphi)^0$ and $s' \in S(\varphi')/S(\varphi')^0$. Define

$$m^{\mu}(\varphi,s;\varphi',s') = \sum_{\epsilon\in\mathcal{E}^{\mu}(\varphi),\epsilon'\in\mathcal{E}^{\mu}(\varphi')} \epsilon(s)\epsilon'(s')m(\varphi,\epsilon;\varphi',\epsilon').$$

By elementary Fourier's inversion, to compute all terms $m(\varphi,\epsilon;\varphi',\epsilon')$ is equivalent to compute all terms $m^{\mu}(\varphi,s;\varphi',s')$. Let d_1,d_2,d'_1,d'_2 be four integers such that d_1 and d_2 are even and $d_1+d_2 = d$, d'_1 and d'_2 are odd and $d'_1+d'_2 = d'+1$, and let δ_1, δ_2 be two elements of $F^{\times}/F^{\times,2}$ such that $\delta_1\delta_2 = \delta$ and, if $d_i = 0$ for $i = 1$ or 2, then $\delta_i = 1$. For $i = 1$ or 2, let $\varphi_i \in \Phi^{orth}_{temp}(d_i,\delta_i)$ and $\varphi'_i \in \Phi^{orth}_{temp}(d'_i)$. To (φ_1,φ_2), we associate as in 4.7 an element $\varphi \in \Phi^{orth}_{temp}(d,\delta)$ and an element $s \in S(\varphi)/S(\varphi)^0$ (as above, we need to choose some additional data, but this choice is inessential). To (φ'_1,φ'_2), we associate as in 4.6 an element $\varphi' \in \Phi^{orth}_{temp}(d')$ and an element $s' \in S(\varphi')/S(\varphi')^0$. The application that, to the data $d_1,d_2,\ldots,\varphi'_2$, associates the data φ, s, φ', s' is surjective (not injective, but this doesn't matter). Then, in order to compute $m^{\mu}(\varphi,s;\varphi',s')$, we can assume that φ, s, ... are deduced from fixed data $d_1, d_2, \ldots, \varphi'_2$. For $i, j \in \{1, 2\}$, Define

$$S(\varphi_i, \varphi'_j) = 2m^+(\varphi_i, 1_{\varphi_i}; \varphi'_j, 1_{\varphi'_j}) - 1.$$

Recall we have introduced some scalars $\gamma^\mu(\varphi_1, \varphi_2)$ and $\gamma^\mu(\varphi'_1, \varphi'_2)$ in 4.6 and 4.7.

PROPOSITION. *We have the equality*

$$\gamma^\mu(\varphi_1, \varphi_2)\gamma^\mu(\varphi'_1, \varphi'_2)m^\mu(\varphi, s; \varphi', s')$$
$$= \frac{1}{2}(S(\varphi_1, \varphi'_1)S(\varphi_2, \varphi'_2) + \mu S(\varphi_1, \varphi'_2)S(\varphi_2, \varphi'_1)).$$

SKETCH OF THE PROOF. We associate to φ and φ' some groups G and G' as above. We associate to d_1, d_2, δ_1 and δ_2 the groups G_1 and G_2 of 4.7 and to d'_1 and d'_2 the groups G'_1 and G'_2 of 4.6. The term $m^\mu(\varphi, s; \varphi', s')$ is a weighted sum of multiplicities that can be computed by the integral formula of 2.5. Then the whole term $m^\mu(\varphi, s; \varphi', s')$ is a sum over tori of integrals of functions, that are weighted sums of the functions appearing in 2.5. In fact, those weighted sums are

$$\sum_{\epsilon \in \mathcal{E}^\mu(\varphi)} \epsilon(s) c_{\pi(\varphi, \epsilon)},$$

and

$$\sum_{\epsilon' \in \mathcal{E}^\mu(\varphi')} \epsilon'(s') c_{\pi(\varphi', \epsilon')}.$$

They can be deduced from the characters $\theta^G(\varphi, s)$ and $\theta^{G'}(\varphi', s')$ introduced in 4.6 and 4.7. Using the conjectures of these sections, they can be computed in terms of the characters $\theta^{G_1}(\varphi_1, 1_{\varphi_1}) \times \theta^{G_2}(\varphi_2, 1_{\varphi_2})$ and $\theta^{G'_1}(\varphi'_1, 1_{\varphi'_1}) \times \theta^{G'_2}(\varphi'_2, 1_{\varphi'_2})$ (here we need to slip the scalars $\gamma^\mu(\varphi_1, \varphi_2)$ and $\gamma^\mu(\varphi'_1, \varphi'_2)$). We obtain an integral formula computing the left member of the proposition, using those characters. The same thing holds for the right member, applying the same methods to the four pairs of groups (G_1, G'_1), (G_2, G'_2), (G_1, G'_2) and (G_2, G'_1). The proposition results from comparison of the two integral formulas obtained in this way

REMARK. Stictly speaking, it is not always possible to attach to G_1, G_2, G'_1 and G'_2 spaces V_1, V_2, V'_1, V'_2, provided with quadratic forms, such that the assumptions of 2.1 are satisfied for the four pairs (V_i, V'_j), $i, j = 1, 2$. That is, we cannot use the same ν for the four pairs. But a term $S(\varphi_i, \varphi'_j)$ depends only on the L-packets $\Pi^{G_i}(\varphi_i)$ and $\Pi^{G'_j}(\varphi'_j)$ and not on the parametrizations of those packets. That is, it does not depend on ν.

We must to compute the terms $S(\varphi_i, \varphi'_j)$. For that, we can forget the indices i and j and to compute $S(\varphi, \varphi')$ defined as above by

$$S(\varphi, \varphi') = 2m^+(\varphi, 1_\varphi; \varphi', 1_{\varphi'}) - 1.$$

So $\mu = 1$ and the underlying groups are quasi-split. We introduce as in 4.6 and 4.7 the representations $\boldsymbol{\pi}(\boldsymbol{\varphi}'_>)$ of $GL(d', F)$ and $\boldsymbol{\pi}(\boldsymbol{\varphi})$ of $GL(d, F)$, and the scalars $c_>(\varphi')$ and $c(\varphi)$. Recall the scalar $\zeta(\varphi)$ of the first assertion of proposition 4.7 (that is proved directly).

PROPOSITION. *We have the equality*

$$S(\varphi, \varphi') = \zeta(\varphi)c(\varphi)c_>(\varphi')\epsilon(\boldsymbol{\pi}(\boldsymbol{\varphi}) \times \boldsymbol{\pi}(\boldsymbol{\varphi}'_>), 1/2, \psi_F).$$

SKETCH OF THE PROOF. We associate to φ and φ' the same groups G and G' as above. The left member $S(\varphi, \varphi')$ can again be computed by an integral formula where appear functions deduced from the characters $\theta^G(\varphi, 1_\varphi)$ and $\theta^{G'}(\varphi', 1_{\varphi'})$. As in 4.5, we associate to G a group $\mathbf{G}$ and a twisted space $\tilde{\mathbf{G}}$, then to φ the function $\theta_{\tilde{\boldsymbol{\pi}}(\boldsymbol{\varphi})}$ on $\tilde{\mathbf{G}}(F)$. As in 4.4, we associate to G' a group $\mathbf{G}'$ and a twisted space $\tilde{\mathbf{G}}'$, then to φ' the function $\theta_{\tilde{\boldsymbol{\pi}}(\boldsymbol{\varphi}'_>)}$ on $\tilde{\mathbf{G}}'(F)$. Using the conjectures of 4.6 and 4.7, we can replace the characters $\theta^G(\varphi, 1_\varphi)$ and $\theta^{G'}(\varphi', 1_{\varphi'})$ by $c(\varphi)\theta_{\tilde{\boldsymbol{\pi}}(\boldsymbol{\varphi})}$ and $c_>(\varphi')\theta_{\tilde{\boldsymbol{\pi}}(\boldsymbol{\varphi}'_>)}$. Then $S(\varphi, \varphi')$ is computed by an integral formula where appear functions deduced from these last characters. Using the proposition 3.4, we see that the same is true for the right member of the proposition. The equality of the two members results from comparison of the two integral formulas obtained in this way.

REMARK. In a first step, this proof gives a more complicated formula because an auxiliary parameter $\boldsymbol{\nu}$ appears in proposition 3.4. But we can choose arbitrarily this parameter. Because the final formula cannot depend on this choice, this implies some relations between the various characters appearing in the formula and these relations can be used to simplify the formula as in the above statement.

Using the two preceding propositions, we obtain an equality between $m^\mu(\varphi, s; \varphi', s')$ and an expression that contains some ϵ-factors and various scalars as $\gamma^\mu(\varphi_1, \varphi_2)$, $c(\varphi_1)$, etc... To determine the scalars, we consider particular cases where some terms can be computed. For instance, assume $d = 0$, $d'_2 = 1$ and $\mu = +1$. Then all terms depending on φ, φ_1, φ_2 and φ'_2 disappears. The only scalars that appear are $c_>(\varphi')$ and $\gamma^+(\varphi', -)$ (the term φ'_2 disappears and $\varphi' = \varphi'_1$). The transfer from $G'_1 = G'$ to G' is the identity and this implies that $\gamma^+(\varphi', -) = 1$. The ϵ-factors are associate to pairs of representations of linear groups, where at least one of the groups is $GL(0)$. Then these ϵ-factors are equal to 1. Our equality reduces to

$$m^+(-, -; \varphi', 1_{\varphi'}) = c_>(\varphi').$$

But, by definition, the left member is a true sum of multiplicities, then is a natural integer. And the right member is a complex number of absolute value 1. Then the two members are equal to 1, proving $c_>(\varphi') = 1$. More complicated particular cases can be used to determine the others scalars. After computation of these scalars, our formula becomes and equality expressing $m^{\mu}(\varphi, s; \varphi', s')$ in terms of ϵ-factors and some computations prove that it is equivalent to the formula of the theorem 5.4.

REMARK. As above, we can apply the constructions to the cases $d = 0$ or $d' = 1$. We obtain the following results. Assume $\mu(G') = 1$ and denote by ϵ'_1 the identity element of $(S(\varphi')/S(\varphi')^0)^\vee$. Then the representation $\pi(\varphi', \epsilon'_1)$ is the unique element of $\Pi^{G'}(\varphi')$ that is generic (cf. 6.2 for the definition). Assume $\mu(G) = 1$ and denote by ϵ_1 the identity element of $(S(\varphi)/S(\varphi)^0)^\vee$. Then the representation $\pi(\varphi, \epsilon_1)$ is the unique element of $\Pi^G(\varphi)$ that is generic, for a type of Whittaker model depending on ν.

6. The case of generic packets

6.1. Definition of non-tempered packets. Let V, q, G be as in 4.4. Let $\varphi \in \Phi^{orth}(d, \delta)$. We don't assume φ tempered. There exists a Levi subgroup $\hat{M}$ of $SO(d, \mathbb{C})$, in the sense of 2.7, an homomorphism $\varphi_{temp}: W_{DF} \to \hat{M}$ that is tempered ($\varphi_{temp}(W_F)$ is relatively compact), and an homomorphism $\hat{\chi}$ from W_{DF} to the center $Z(\hat{M})$, such that:

- the homomorphism $\hat{\chi}$ is positive; that means that for every rational character $x^*: Z(\hat{M}) \to \mathbb{C}^\times$, $x^* \circ \hat{\chi}$ takes real positive values; this implies $\hat{\chi}$ is in fact defined on the quotient W_F/I_F of W_{DF};
- the homomorphism φ is equivalent to $\hat{\chi} \otimes \varphi_{temp}$;
- if another triple $\hat{M}_1$, $\varphi_{temp,1}$ and $\hat{\chi}_1$ satisfy the same properties, then $\hat{M}_1 \subset \hat{M}$.

In general, the group $\hat{M}$ defines by duality a Levi subgroup M of G. It is not true in some particular case. We ignore temporarily this difficulty. The group M is a product of groups $GL(n)$ and (possibly) one special orthogonal group of the even-dimensional case. We can apply to this last group the conjecture of 4.7 and to the components $GL(n)$ the known correspondence of Langlands. Then the homomorphism φ_{temp} defines a packet $\Pi^M(\varphi_{temp})$ of smooth irreducible tempered representations of $M(F)$. The character $\hat{\chi}$ defines a character χ^M of $M(F)$. We can choose a parabolic subgroup P of G, of Levi component M, such that $\hat{\chi}$ satisfies relatively to P the positivity condition required for applying the Langlands quotient theorem. Then this theorem says that, for all $\tau \in \Pi^M(\varphi_{temp})$, the induced representation $Ind_P^G(\chi^M \otimes \tau)$ has a unique irreducible quotient. Denote $\mathcal{L}(\chi^M \otimes \tau)$ this quotient. By definition, the L-packet $\Pi^G(\varphi)$ is the set of such representations $\mathcal{L}(\chi^M \otimes \tau)$, when τ describes $\Pi^M(\varphi_{temp})$.

Let's come back to the particular case where M does not exist. This case happens when G is not quasi-split and $\hat{M}$ is a product of groups $GL(n, \mathbb{C})$ only, without a special orthogonal group. In this case, we define $\Pi^G(\varphi) = \emptyset$.

We see easily that $S(\varphi)$ is equal to the centralizer of the image of φ_{temp} in $\hat{M}$. Then $\Pi^G(\varphi)$ is still parametrized by $\mathcal{E}^{\mu(G)}(\varphi)$.

Now let V', q', G' be as in 4.3 and let $\varphi' \in \Phi^{orth}(d')$. We define the packet $\Pi^{G'}(\varphi')$ by a similar construction.

6.2. Definition of generic packets. We consider two sets of data V'_i, q'_i, G'_i and V'_a, q'_a, G'_a as in 4.3, of the same odd dimension d' and the same discriminant, and we assume $\mu(G'_i) = 1$, $\mu(G'_a) = -1$ (in other words, G'_i is split and G'_a is not-split). Let $\varphi' \in \Phi^{orth}(d')$. The L-packet $\Pi^{G'_i}(\varphi')$ is parametrized by $\mathcal{E}^+(\varphi')$ and the L-packet $\Pi^{G'_a}(\varphi')$ is parametrized by $\mathcal{E}^-(\varphi')$. As in 1.3, let $\Pi(\varphi') = \Pi^{G'_i}(\varphi') \sqcup \Pi^{G'_a}(\varphi')$.

Because G'_i is split, we can fix a Borel subgroup B' of G'_i, defined over F. Denote by U' the unipotent radical of B'. Fix a character ψ of $U'(F)$. We assume that, for all parabolic subgroup P' of G'_i defined over F such that $B' \subset P' \neq G'_i$, ψ is not trivial on $U^{P'}(F)$, where $U^{P'}$ is the unipotent radical of P'. Define the space $\mathcal{W}_\psi$ of locally constant functions $f : G'_i(F) \to \mathbb{C}$ such that $f(ug) = \psi(u)f(g)$ for all $u \in U'(F)$ and $g \in G'_i(F)$ and such that $|f|$ is of compact support on $U'(F)\backslash G'_i(F)$. The group $G'_i(F)$ acts by right translations on this space. For a smooth irreducible representation π' of $G'_i(F)$, we say π' is generic if it is equivalent to a representation in a subspace of $\mathcal{W}_\psi$. This notion does not depend on ψ because all ψ's are conjugate by the action of $B'(F)$. We say that $\Pi(\varphi')$ is generic if $\Pi^{G'_i}(\varphi')$ contains a generic representation.

Now, we consider two sets of data V_i, q_i, G_i and V_a, q_a, G_a as in 4.4, of the same even dimension d and the same discriminant δ, and we assume $\mu(G_i) = 1$, $\mu(G_a) = -1$ (the parameter ν used in the definition of these signs being always fixed). The group G'_i is quasi-split. Let $\varphi \in \Phi^{orth}(d, \delta)$ and let $\Pi(\varphi) = \Pi^{G_i}(\varphi) \sqcup \Pi^{G_a}(\varphi)$. Suppressing the $'$, we define as before the notion of genericity of a smooth irreducible representation of $G_i(F)$ or of the packet $\Pi(\varphi)$. Here, there is a problem because the possible characters ψ are not all conjugate by the action of $B(F)$. A representation of $G_i(F)$ can be generic for one choice of ψ and not generic for another choice. But there is no problem for the packets: we see that, if the packet $\Pi^{G_i}(\varphi)$ contains a representation generic for one choice, then for any choice, it contains a generic representation.

6.3. The theorem for generic L-packets. We consider the same data as in 5.4, but we remove the assumption that φ and φ' are tempered. So we have only $\varphi \in \Phi^{orth}(d, \delta)$ and $\varphi' \in \Phi^{orth}(d')$. Using the definitions of 6.1, the statement of theorem 5.5 has still a meaning. Without additional assumptions, this theorem is false.

THEOREM (joint with Moeglin). *The assertions of theorem 5.5 become true if* $\Pi(\varphi)$ *and* $\Pi(\varphi')$ *are generic.*

That is the theorem C of 1.6. Note that a tempered L-packet is generic: as explained in the last remark of 5.5, that follows from theorem 5.5 applied to the cases $d = 0$ or $d' = 1$. So the preceding theorem is stronger as theorem 5.5.

I give a sketch of the proof. To φ, we associate as in 6.1 an homomorphism φ_{temp} and, for an index $\sharp = i$ or a, a Levi subgroup $M_\sharp$, a parabolic subgroup $P_\sharp$ of $G_\sharp$ and a character $\chi^{M_\sharp}$ of $M_\sharp(F)$ (assuming that $M_\sharp$ exists). Similarly, to φ', we associate φ'_{temp}, $M'_\sharp$, $P'_\sharp$ and $\chi^{M'_\sharp}$.

We have a first result. Let $\tau \in \Pi^{M_\sharp}(\varphi_{temp})$ and $\tau' \in \Pi^{M'_\sharp}(\varphi'_{temp})$. Assume the representations $Ind_{P_\sharp}^{G_\sharp}(\chi^{M_\sharp} \otimes \tau)$ and $Ind_{P'_\sharp}^{G'_\sharp}(\chi^{M'_\sharp} \otimes \tau')$ are irreducible. Then

$$m(Ind_{P_\sharp}^{G_\sharp}(\chi^{M_\sharp} \otimes \tau), Ind_{P'_\sharp}^{G'_\sharp}(\chi^{M'_\sharp} \otimes \tau')) = m(\tau, \tau'),$$

the last multiplicity being defined as in 2.7. This comes easily, using the proposition 2.8.

The second ingredient is more subtle. We fix a character ψ as in 6.2. Because φ_{temp} is tempered, there is a unique representation in $\Pi^{M_i}(\varphi_{temp})$ that is generic. Denote it by τ_{Wh}.

THEOREM. *The three conditions are equivalent:*

(i) $\Pi(\varphi)$ *is generic;*

(ii) *the representation* $Ind_{P_i}^{G_i}(\chi^{M_i} \otimes \tau_{Wh})$ *is irreducible;*

(iii) *for* $\sharp = i$ *or* a *and for all* $\tau \in \Pi^{M_\sharp}(\varphi_{temp})$, *the representation* $Ind_{P_\sharp}^{G_\sharp}(\chi^{M_\sharp} \otimes \tau)$ *is irreducible.*

Cf. [MW] corollary 2.14. The equivalence between (i) and (ii) was a conjecture of Shahidi, and was proved by Heiermann and Muic ([HM]). The new point is the equivalence with (iii). The rather winding proof uses the fine study by Moeglin of the relations between Jacquet modules and endoscopy.

A similar result holds for the groups $G'_\sharp$.

Now, assume $\Pi(\varphi)$ and $\Pi(\varphi')$ are generic. By the two results above, we have the equality

$$m(\mathcal{L}(\chi^{M_\sharp} \otimes \tau), \mathcal{L}(\chi^{M'_\sharp} \otimes \tau')) = m(\tau, \tau')$$

for $\sharp = i$ or a and for all $\tau \in \Pi^{M_\sharp}(\varphi_{temp})$ and $\tau' \in \Pi^{M'_\sharp}(\varphi'_{temp})$. Then the theorem above comes from the theorem 5.5 of the tempered case.

References

[AGRS] A. Aizenbud, D. Gourevitch, S. Rallis, G. Schiffmann: *Multiplicity one theorems*, Annals of Math. 172 (2010), p. 1407–1434

[A] J. Arthur: *Endoscopic classification of representations: orthogonal and symplectic groups*, in preparation

[GGP1] W.T. Gan, B. Gross, D. Prasad: *Restrictions of representations of classical groups: examples*, preprint 2009

[GGP2] ———: *Symplectic local root numbers, central critical L-values, and restriction problems in the representation theory of classical groups*, preprint 2009

[GP] B. Gross, D. Prasad: *On the decomposition of a representation of SO_n when restricted to SO_{n-1}*, Can. J. of Math. 44 (1992), p.974–1002

[GR] B. Gross, M. Reeder: *From Laplace to Langlands via representations of orthogonal groups*, Bull. AMS 43 (2006), p.163–205

[HC] Harish-Chandra: *Admissible invariant distributions on reductive p-adic groups*, notes by S. DeBacker and P. Sally, University lecture series 16, AMS 1999

[HT] M. Harris, R. Taylor: *On the geometry and cohomology of some simple Shimura varieties*, Ann. of Math. Studies 151, PUP 2002

[HM] V. Heiermann, G. Muic: *The standard modules conjecture*, Math. Zeit. 255 (2007), p.19–37

[H] G. Henniart: *Une preuve simple des conjectures de Langlands pour $GL(n)$ sur un corps p-adique*, Invent. Math. 139 (2000), p.439–456

[II] A. Ichino, T. Ikeda: *On the periods of automorphic forms on special orthogonal groups and the Gross-Prasad conjecture*, preprint

[JPSS] H. Jacquet, I. Piatetski-Shapiro, J. Shalika: *Rankin-Selberg convolutions*, American J. of Math. 105 (1983), p.367–464

[MW] C. Moeglin, J.-L. Waldspurger: *La conjecture locale de Gross-Prasad pour les groupes spéciaux orthogonaux: le cas général*, preprint 2009

[W1] J.-L. Waldspurger: *Une formule intégrale reliée à la conjecture locale de Gross-Prasad*, Compositio Math. 146 (2010), p.1180–1290

[W2] ———: *Une formule intégrale reliée à la conjecture locale de Gross-Prasad, 2ème partie: extension aux représentations tempérées*, to appear in Astérisque

[W3] ———: *Calcul d'une valeur d'un facteur ϵ par une formule intégrale*, preprint 2009

[W4] ———: *La conjecture locale de Gross-Prasad pour les représentations tempérées des groupes spéciaux orthogonaux*, preprint 2009

Current Developments in Mathematics, 2010

Wellposedness of the two- and three-dimensional full water wave problem

Sijue Wu

1. Introduction

The mathematical problem of n-dimensional water wave concerns the motion of the interface separating an inviscid, incompressible, irrotational fluid, under the influence of gravity, from a region of zero density (i.e. air) in n-dimensional space. It is assumed that the fluid region is below the air region. Assume that the density of the fluid is 1, the gravitational field is $-\mathbf{k}$, where $\mathbf{k}$ is the unit vector pointing in the upward vertical direction, and at time $t \geq 0$, the free interface is $\Sigma(t)$, and the fluid occupies region $\Omega(t)$. When surface tension is zero, the motion of the fluid is described by

$$(1.1) \qquad \begin{cases} \mathbf{v}_t + \mathbf{v} \cdot \nabla \mathbf{v} = -\mathbf{k} - \nabla P & \text{on } \Omega(t), \ t \geq 0, \\ \operatorname{div} \mathbf{v} = 0, \quad \operatorname{curl} \mathbf{v} = 0, & \text{on } \Omega(t), \ t \geq 0, \\ P = 0, & \text{on } \Sigma(t) \\ (1, \mathbf{v}) \text{ is tangent to the free surface } (t, \Sigma(t)), \end{cases}$$

where $\mathbf{v}$ is the fluid velocity, P is the fluid pressure. It is well-known that when surface tension is neglected, the water wave motion can be subject to the Taylor instability [**30, 3, 2**]. Assume that the free interface $\Sigma(t)$ is described by $\xi = \xi(\alpha, t)$, where $\alpha \in R^{n-1}$ is the Lagrangian coordinate, i.e. $\xi_t(\alpha, t) = \mathbf{v}(\xi(\alpha, t), t)$ is the fluid velocity on the interface, $\xi_{tt}(\alpha, t) = (\mathbf{v}_t + \mathbf{v} \cdot \nabla \mathbf{v})(\xi(\alpha, t), t)$ is the acceleration. Let $\mathbf{n}$ be the unit normal pointing out of $\Omega(t)$. The Taylor sign condition relating to Taylor instability is

$$(1.2) \qquad -\frac{\partial P}{\partial \mathbf{n}} = (\xi_{tt} + \mathbf{k}) \cdot \mathbf{n} \geq c_0 > 0,$$

point-wisely on the interface for some positive constant c_0. In [**32, 33**], we showed that the Taylor sign condition (1.2) always holds for the n-dimensional infinite depth water wave problem (1.1), $n \geq 2$, as long as the interface is non-self-intersecting; and the initial value problem of the

Financial support in part by NSF grant DMS-0800194.

water wave system (1.1) is uniquely solvable locally in time in Sobolev spaces for arbitrary given data. Earlier work include Nalimov [**20**], Yosihara [**36**] and Craig [**11**] on local existence and uniqueness for small data in 2D. Notice that if surface tension is not zero, or if there is a bottom, or nonzero-vorticity, the Taylor sign condition need not hold. Local wellposedness for water wave motion with the effect of surface tension, bottom and a non-zero vorticity, under the assumption (1.2) can be found in [**1, 5, 6, 15, 18, 19, 21, 25, 37**].

To understand the large time behavior of solutions for the water wave problem (1.1), it is necessary to understand the nature of its nonlinearity.

From the fluids literature [**12**], we know that "wavelengths which are generated in the open ocean can be very large; swell as long as 800m between crests has been recorded. The wave heights are limited, the theoretical limit being about one seventh of the wave length. Records of giant waves have to be regarded with some caution, and a survey taken from sailing ships' logs showed that 45% of waves are less than 1.25m in height between trough and crest, with only 10% greater than 6m. Nevertheless there is a reliable record of a wave 34m in height.

We shall see that waves of different wavelengths travel at different speeds, so that a giant wave maybe produced for a short time by a combination of shorter and longer waves. However, in some parts of the world the occurrence of giant waves is apparently more frequent, for example off the east coast of South Africa, where long waves originating in the 'roaring forties' meet the Agulhas current flowing towards them. Adverse currents amplify waves and can bend them to give a focusing effect. There are numerous reports of ships being damaged by waves in this area. In 1973 two large ships were damaged.... Something similar probably accounted for the disappearance of the Waratah, a liner on only her second voyage, in 1909. No trace of her was ever found, although there had been other ships in the area at the time, including sailing ships, which had not been in difficulties."

Such giant waves is the so called rogue waves. From the above description it seems that rogue waves can appear in perfectly clear weather in open ocean. Conjectures on possible causes include: diffractive focusing due to certain coastal shapes; focusing by currents; and nonlinear effect–such as that of a focusing nonlinear Schrödinger equation.

In our recent work [**34, 35**] we focused on understanding the nonlinear effect. We therefore assumed there is no wind, no coast, no bottom. For large scale waves such as those in open ocean, it is reasonable to neglect surface tension. We studied the behavior of solutions of system (1.1) for large time for data that are small in appropriate senses.

In [**34, 35**], we showed that for the 2D water wave problem (1.1) ($n = 2$), the quantity $\Theta = (\bar{I} - \mathfrak{H})y$, under an appropriate coordinate change $k = k(\alpha, t)$, satisfies an equation of the type

$$\partial_t^2\Theta - i\partial_\alpha\Theta = \mathcal{G} \tag{1.3}$$

with $\mathcal{G}$ consisting of nonlinear terms of only cubic and higher orders. Here $\mathfrak{H}$ is the Hilbert transform related to the water region $\Omega(t)$, y is the height function for the interface $\Sigma(t) : \xi = (x(\alpha, t), y(\alpha, t)), \alpha \in \mathbb{R}$. Similarly, for the 3D water wave problem (1.1) ($n = 3$), if $\Sigma(t) : \xi = (x(\alpha, \beta, t), y(\alpha, \beta, t), z(\alpha, \beta, t))$, $(\alpha, \beta) \in \mathbb{R}^2$ is the interface at time t, and $\mathfrak{H}$ is the Hilbert transform associated to the water region $\Omega(t)$, $N = \xi_\alpha \times \xi_\beta$ is the outward normal, then the quantity $\theta = (I - \mathfrak{H})z$ satisfies such an equation

$$\partial_t^2 \theta - \mathfrak{a} N \times \nabla \theta = G \tag{1.4}$$

where G is a nonlinearity of cubic and higher orders in nature. Furthermore we found a coordinate change k that transforms (1.4) into an equation consisting of a linear part plus only cubic and higher order nonlinear terms. Using these favorable structures, we showed that for water waves in two space dimensions, if initially the amplitude of the interface (and finitely many of its derivatives) is of size $O(\epsilon)$ and the kinetic energy is of size $O(\epsilon^2)$ and small, then there exists a unique classical solution of the water wave equation (1.1) ($n = 2$) for a time period $[0, e^{c/\epsilon}]$; during this time period, the amplitude of the interface remains small and is as regular as the initial interface. Here c is a constant independent of ϵ (c.f. Theorem 3.8 and [**34**]). For water waves in three space dimensions, if initially the steepness of the interface and the fluid velocity on the interface (and finitely many of their derivatives) are small, then there exists a unique classical solution of the water wave equation (1.1) ($n = 3$) for all time, and the interface remains to have small steepness and is as regular as the initial interface for all time (c.f. Theorem 3.9 and [**35**]). We note that in our result for the 3D water waves, no direct assumptions are made to the height of the initial interface and the velocity field in the fluid domain. In particular, the amplitude of the initial interface can be arbitrary large, the initial kinetic energy $\frac{1}{2}\|\mathbf{v}\|^2_{L^2(\Omega(0))}$ can be infinite. We also note that although our result for the 2D water waves [**34**] requires the initial kinetic energy $\frac{1}{2}\|\mathbf{v}\|^2_{L^2(\Omega(0))}$ being small, while viewed as a special case of 3D where the wave is constant in one direction, its 3D energy in fact is infinite ($\infty \times \epsilon^2 = \infty$).

Before moving on to some detailed discussions, we mention the main idea in proving the Taylor sign condition (1.2) for the water wave equation (1.1):

Applying div to both sides of the Euler equation, and using the assumption that $\operatorname{curl}\mathbf{v} = 0$, we find

$$\Delta P = -|\nabla \mathbf{v}|^2 \leq 0 \quad \text{in } \Omega(t).$$

Therefore from $P = 0$ on the interface $\Sigma(t)$ and the maximal principle, we have $-\frac{\partial P}{\partial \mathbf{n}} \geq 0$. For a precise proof of the fact that the Taylor sign condition (1.2) always holds, we refer to [**32, 33**].

We now give some details on what we have obtained so far (c.f. [**32, 33, 34, 35**]). Notice that equation (1.1) is a nonlinear equation defined in moving domains. It is difficult to obtain results directly from it. One key step in our approach is to rewrite (1.1) into forms from which results and informations can be obtained.

2. Local wellposedness of the 2D and 3D water wave problems

We first discuss some of the main ideas in proving the local in time well-posedness of the water wave system (1.1) [**32, 33**]. For clarity, we mainly write in terms of the 2D water wave, we will only mention the main differences and ideas for 3D.

Let $\xi = \xi(\alpha, t) = (x(\alpha, t), y(\alpha, t))$, $\alpha \in \mathbb{R}$, be the equation of the free interface $\Sigma(t)$ at time t in Lagrangian coordinates α. In what follows, we regard the 2-D space as a complex plane and use the same notation for a complex form $\xi = x + i\,y$ and a point $\xi = (x, y)$. So $\bar{\xi} = x - i\,y$. We write $[A, B] = AB - BA$.

As we know it is difficult to solve (1.1) directly since it is defined on moving domains. The first step used in [**32, 33**] in solving (1.1) was to reduce it to an equivalent equation defined on the moving interface $\Sigma(t)$.[1] This was done based on the following observations. First notice that the second equation in (1.1): $\operatorname{div} \mathbf{v} = 0$, $\operatorname{curl} \mathbf{v} = 0$ implies that the complex conjugate of the velocity field $\bar{\mathbf{v}}$ is holomorphic in $\Omega(t)$. Therefore the trace of the velocity on $\Sigma(t)$: $\xi_t(\cdot, t) = \mathbf{v}(\xi(\cdot, t), t)$ satisfies $\bar{\xi}_t = \mathfrak{H}\bar{\xi}_t$, where

$$\mathfrak{H} f(\alpha, t) = \frac{1}{\pi i} p.v. \int \frac{f(\beta, t)\xi_\beta(\beta, t)}{\xi(\alpha, t) - \xi(\beta, t)}\, d\beta \tag{2.1}$$

is the Hilbert transform on $\Sigma(t)$ associated with the parameterization $\xi = \xi(\alpha, t)$. Now from the third equation in (1.1): $P = 0$ on $\Sigma(t)$, we know ∇P points in the direction normal to $\Sigma(t)$, therefore $-\nabla P = i\mathfrak{a}\xi_\alpha$, with $\mathfrak{a} = -\frac{\partial P}{\partial \mathbf{n}}\frac{1}{|\xi_\alpha|}$. So the trace of the Euler equation on $\Sigma(t)$ can be written as $\xi_{tt} + i = i\mathfrak{a}\xi_\alpha$, and the system (1.1) is equivalent to the following system on the interface $\Sigma(t)$:

$$\begin{cases} \xi_{tt} + i = i\mathfrak{a}\xi_\alpha \\ \bar{\xi}_t = \mathfrak{H}\bar{\xi}_t \end{cases} \tag{2.2}$$

However (2.2) is fully nonlinear. To solve (2.2), we further reduced it into a quasilinear equation [**32, 33**], so that on which the classical energy method could be applied. This was done by taking one derivative to t to the first equation in (2.2). Taking derivative to t to the first equation in (2.2), we got

$$\begin{cases} \xi_{ttt} - i\mathfrak{a}\xi_{t\alpha} = i\mathfrak{a}_t\xi_\alpha \\ \bar{\xi}_t = \mathfrak{H}\bar{\xi}_t \end{cases} \tag{2.3}$$

[1]The approaches of Nalimov [**20**], Yosihara [**36**] and Craig [**11**] were also to reduce the system (1.1), for $n = 2$ only, to a system defined on the moving interface.

Using the fact that $\bar{\xi}_t = \mathfrak{H}\bar{\xi}_t$, we deduced

$$\begin{aligned}(I-\mathfrak{H})(-i\mathfrak{a}_t\bar{\xi}_\alpha) &= (I-\mathfrak{H})(\bar{\xi}_{ttt} + i\mathfrak{a}\bar{\xi}_{t\alpha}) \\ &= [\partial_t^2 + i\mathfrak{a}\partial_\alpha, \mathfrak{H}]\bar{\xi}_t \\ &= 2[\xi_{tt}, \mathfrak{H}]\frac{\bar{\xi}_{t\alpha}}{\xi_\alpha} + 2[\xi_t, \mathfrak{H}]\frac{\bar{\xi}_{tt\alpha}}{\xi_\alpha} - \frac{1}{\pi i}\int \left(\frac{\xi_t(\alpha,t)-\xi_t(\beta,t)}{\xi(\alpha,t)-\xi(\beta,t)}\right)^2 \bar{\xi}_{t\beta}\, d\beta \end{aligned} \tag{2.4}$$

Further using the fact that $\mathfrak{a}$ and $\mathfrak{a}_t$ are real valued gave

$$(I+\mathfrak{K}^*)(\mathfrak{a}_t|\xi_\alpha|) = Re\left(\frac{i\xi_\alpha}{|\xi_\alpha|}\left\{2[\xi_{tt}, \mathfrak{H}]\frac{\bar{\xi}_{t\alpha}}{\xi_\alpha} + 2[\xi_t, \mathfrak{H}]\frac{\bar{\xi}_{tt\alpha}}{\xi_\alpha} - \frac{1}{\pi i}\int \left(\frac{\xi_t(\alpha,t)-\xi_t(\beta,t)}{\xi(\alpha,t)-\xi(\beta,t)}\right)^2 \bar{\xi}_{t\beta}\, d\beta\right\}\right) \tag{2.5}$$

here $Re\xi$ indicates the real part of ξ,

$$\mathfrak{K}^* f(\alpha,t) = p.v. \int Re\left\{\frac{-1}{\pi i}\frac{\xi_\alpha}{|\xi_\alpha|}\frac{|\xi_\beta(\beta,t)|}{(\xi(\alpha,t)-\xi(\beta,t))}\right\} f(\beta,t)\, d\beta$$

is the adjoint of the double layered potential operator $\mathfrak{K}$ in $L^2(\Sigma(t), dS)$. Notice that $I + \mathfrak{K}^*$ is invertible on $L^2(\Sigma(t), dS)$. From (2.5) we see that $\mathfrak{a}_t|\xi_\alpha|$ has the same regularity as that of $\bar{\xi}_{tt}$ and $\bar{\xi}_t$.

We now rewrite (2.3) as

$$\begin{cases} \xi_{ttt} - i\mathfrak{a}\xi_{t\alpha} = \dfrac{\xi_{tt}+i}{|\xi_{tt}+i|}\mathfrak{a}_t|\xi_\alpha| \\ \bar{\xi}_t = \mathfrak{H}\bar{\xi}_t, \end{cases} \tag{2.6}$$

and use (2.5) for $\mathfrak{a}_t|\xi_\alpha|$. Notice that $i\partial_\alpha\xi_t = -\nabla_{\mathbf{n}}\xi_t$, and the Dirichlet-Neumann operator $\nabla_{\mathbf{n}}$ is a positive operator. As we have seen the Taylor sign condition (1.2) holds, i.e. $\mathfrak{a} > 0$ [**32, 33**]. Therefore (2.6) is a weakly hyperbolic quasi-linear system with the right hand side of the first equation in (2.6) consisting of terms of lower order derivatives of ξ_t. The local in time wellposedness of (2.6)-(2.5) in Sobolev spaces (with $(\xi_t, \xi_{tt}) \in C([0,T], H^{s+1/2} \times H^s)$) was then proved by energy estimates and a fixed point iteration argument. Through establishing the equivalence of (1.1) with (2.6)-(2.5), we obtained the local in time well-posedness in Sobolev spaces of the full water wave equation (1.1). For details of the proof, we refer the readers to [**32, 33**].

We now mention some of the main ideas in deriving the quasi-linear system for the 3D water waves. To derive a similar equivalent quasilinear system for the 3D water wave equation (1.1) ($n = 3$), naturally we looked for a suitable counterpart in 3D of the equation for the trace on the interface of "div $\mathbf{v} = 0$, curl $\mathbf{v} = 0$". We found Clifford analysis offers a suitable framework for 3D water waves. Let's recall the basics of Clifford algebra, or in other words, the algebra of quaternions $\mathcal{C}(V_2)$ (c.f. [**14**]). Let $\{1, e_1, e_2, e_3\}$

be the basis of $\mathcal{C}(V_2)$, satisfying

$$e_i^2 = -1, \quad \text{for } i = 1, 2, 3, \quad e_i e_j = -e_j e_i, \;\; i \neq j, \quad e_3 = e_1 e_2. \tag{2.7}$$

Let $\mathcal{D} = \partial_x e_1 + \partial_y e_2 + \partial_z e_3$. By definition, a Clifford-valued function $F : \Omega \subset \mathbb{R}^3 \to \mathcal{C}(V_2)$ is Clifford analytic in domain Ω if $\mathcal{D}F = 0$ in Ω. Therefore $F = \sum_{i=1}^3 f_i e_i$ is Clifford analytic in Ω if and only if $\text{div}F = 0$ and $\text{curl}F = 0$ in Ω. Furthermore we know Clifford analytic functions have the property that F is Clifford analytic in Ω if and only if $F = \mathfrak{H}_\Sigma F$, where

$$\mathfrak{H}_\Sigma g(\alpha, \beta) = p.v. \iint K(\eta(\alpha', \beta') - \eta(\alpha, \beta))\, (\eta'_{\alpha'} \times \eta'_{\beta'})\, g(\alpha', \beta')\, d\alpha' d\beta' \tag{2.8}$$

is the 3D version of the Hilbert transform on $\Sigma = \partial\Omega : \eta = \eta(\alpha, \beta),\ (\alpha, \beta) \in \mathbb{R}^2$, with normal $\eta_\alpha \times \eta_\beta$ pointing out of Ω, and

$$\Gamma(\eta) = -\frac{1}{\omega_3 |\eta|}, \quad K(\eta) = -2\mathcal{D}\Gamma(\eta) = -\frac{2}{\omega_3} \frac{\eta}{|\eta|^3},$$

ω_3 is the surface area of the unit sphere in $\mathbb{R}^3$.

Therefore if $\xi = \xi(\alpha, \beta, t)$, $(\alpha, \beta) \in \mathbb{R}^2$ is the free interface $\Sigma(t)$ in Lagrangian coordinates (α, β) at time t, with $N = \xi_\alpha \times \xi_\beta$ pointing out of the fluid domain, we can rewrite the 3D water wave system (1.1) ($n = 3$) as (c.f. [**33**])

$$\begin{cases} \xi_{tt} + e_3 = \mathfrak{a} N \\ \xi_t = \mathfrak{H}_{\Sigma(t)} \xi_t \end{cases} \tag{2.9}$$

where $\mathfrak{a} = -\frac{1}{|N|}\frac{\partial P}{\partial \mathbf{n}}$. From the nonlinear system (2.9) we derived the quasi-linear system (cf. (5.21)-(5.22) of [**33**]) for the 3D water waves, and the local in time well-posedness of the 3D full water wave system was therefore obtained from energy estimates and a fixed point iteration argument to the quasi-linear system [**33**].

Before ending this section, we point out that the quasilinear system (2.6)-(2.5) (and (5.21)-(5.22) of [**33**] for 3D water waves) is coordinate invariant.

Let $U_g f(\alpha, t) = f \circ g(\alpha, t) = f(g(\alpha, t), t)$. For fixed t, let $k = k(\alpha, t) : \mathbb{R} \to \mathbb{R}$ be a diffeomorphism and $k_\alpha > 0$. Let k^{-1} be such that $k \circ k^{-1}(\alpha, t) = \alpha$. Define

$$\zeta = \xi \circ k^{-1}, \quad b = k_t \circ k^{-1} \quad \text{and} \quad A \circ k = \mathfrak{a} k_\alpha \tag{2.10}$$

Let

$$D_t = U_k^{-1} \partial_t U_k = \partial_t + b\partial_\alpha$$

be the material derivative. By a simple application of the chain rule, we have

$$U_k^{-1}(\partial_t^2 - i\mathfrak{a}\partial_\alpha) U_k = D_t^2 - iA\partial_\alpha,$$

and equation (2.6) becomes

$$(2.11)\qquad \begin{cases} (D_t^2 - iA\partial_\alpha)D_t\zeta = (\mathfrak{a}_t|\xi_\alpha|)\circ k^{-1}\dfrac{D_t^2\zeta + i}{|D_t^2\zeta + i|} \\ \overline{D_t\zeta} = \mathcal{H}\overline{D_t\zeta} \end{cases}$$

with

$$\begin{aligned} &(I+\mathcal{K}^*)((\mathfrak{a}_t|\xi_\alpha|)\circ k^{-1}) \\ &= Re(\frac{i\zeta_\alpha}{|\zeta_\alpha|}\{2[D_t^2\zeta,\mathcal{H}]\frac{\partial_\alpha\overline{D_t\zeta}}{\zeta_\alpha} + 2[D_t\zeta,\mathcal{H}]\frac{\partial_\alpha\overline{D_t^2\zeta}}{\zeta_\alpha} \\ (2.12)\qquad &- \frac{1}{\pi i}\int(\frac{D_t\zeta(\alpha,t) - D_t\zeta(\beta,t)}{\zeta(\alpha,t)-\zeta(\beta,t)})^2\partial_\beta\overline{D_t\zeta}(\beta,t)\,d\beta\}) \end{aligned}$$

and

$$(2.13)\qquad \mathcal{H}f(\alpha,t) = U_k^{-1}\mathfrak{H}U_kf(\alpha,t) = \frac{1}{\pi i}p.v.\int\frac{f(\beta,t)\zeta_\beta(\beta,t)}{\zeta(\alpha,t)-\zeta(\beta,t)}\,d\beta,$$

$$(2.14)\qquad \mathcal{K}^*f(\alpha,t) = p.v.\int Re\{\frac{-1}{\pi i}\frac{\zeta_\alpha}{|\zeta_\alpha|}\frac{|\zeta_\beta(\beta,t)|}{(\zeta(\alpha,t)-\zeta(\beta,t))}\}f(\beta,t)\,d\beta.$$

Notice the remarkable similarities between equations (2.6)-(2.5) and (2.11)-(2.12). In particular, the structures of the terms in (2.6)-(2.5) do not change under the change of variables. This makes it convenient for us to work in another coordinate system and to choose a different coordinate system when there is advantage to do so. In fact, using (2.11)-(2.12) (and the corresponding 3D version) one can prove the local in time well-posedness of the water wave problem in this arbitrarily chosen coordinate system using the same analysis as in [**32, 33**].

We now turn to the question of long time behavior of the solutions for the water wave equation (1.1) for small initial data.

3. Global and almost global well-posedness of the 2D and 3D water wave problem

Our study of the global in time behavior of the water wave equation (1.1) further uses two new ideas, one is the dispersive aspect of the water wave motion, another is a better understanding of the nature of the nonlinearity of (1.1).

Let $\mathfrak{u} = \xi_t$. Linearizing the quasi-linear system (2.6)-(2.5) (or the quasi-linear system (5.21)-(5.22) of [**33**] for 3D water waves) at the zero solution, we obtain

$$(3.1)\qquad \partial_t^2\mathfrak{u} + |D|\mathfrak{u} = F(\mathfrak{u}_t, |D|\mathfrak{u}),\quad (\alpha,t)\in\mathbb{R}^{n-1}\times\mathbb{R}$$

where $|D| = \sqrt{-\Delta}$, Δ is the Laplacian in $\mathbb{R}^{n-1}$ for n-D water waves, F consists of the nonlinear terms. It is not surprising that F contains quadratic nonlinear terms. We know the dispersion relation of the free equation

$$\partial_t^2 \mathfrak{u} + |D|\mathfrak{u} = 0 \tag{3.2}$$

is

$$\omega^2 = |k| \tag{3.3}$$

for plane wave solutions $\mathfrak{u} = e^{i(k\cdot\alpha+\omega t)}$. So waves of wave number k travel with phase velocity $\frac{\omega}{|k|}\hat{k}$, where $\hat{k} = \frac{k}{|k|}$, equation (3.1) is dispersive. We know for a large class $\mathcal{B}$ of smooth initial data, the solution of the linear equation (3.2) exists for all time and remains smooth, and its L^∞ norm decays with rate $1/t^{\frac{n-1}{2}}$. But nonlinear interactions can cause blow-up of the solutions at finite time. So to answer the question that for small data in $\mathcal{B}$, for how long does the solution of the nonlinear equation (3.1) remain smooth, we need to know for how long does the linear part of the equation (3.1) remain dominant. The weaker the nonlinear interaction, the longer the solution remains smooth. For small data, quadratic interactions are in general stronger than the cubic and higher order interactions.

To illustrate this idea, we look at the following model equation with the $(p+1)$th-order nonlinearity:

$$\partial_t^2 \mathfrak{u} + |D|\mathfrak{u} = (\partial_t \mathfrak{u})^{p+1}, \quad (\alpha, t) \in \mathbb{R}^{n-1} \times \mathbb{R} \tag{3.4}$$

Suppose we can prove such decay estimates for the solution: for $i \le s - 10$,

$$|\partial^i \partial_t \mathfrak{u}(t)|_{L^\infty} \lesssim (1+t)^{-\frac{n-1}{2}} E_s(t)^{1/2}$$

where ∂ is some kind of derivatives,

$$E_s(t) = \sum_{|j|\le s} \int |\partial^j \partial_t \mathfrak{u}(\alpha, t)|^2 + |\partial^j |D|^{1/2} \mathfrak{u}(\alpha, t)|^2 \, d\alpha$$

Then we can derive energy estimates for $s \ge 20$,

$$\frac{d}{dt} E_s(t) \lesssim (1+t)^{-\frac{(n-1)p}{2}} E_s(t)^{p/2+1}$$

therefore

$$E_s(0)^{-p/2} - E_s(T)^{-p/2} \lesssim \int_0^T (1+t)^{-\frac{(n-1)p}{2}} \, dt$$

Heuristically, we expect to prove existence of solutions of (3.4) for as long as the energy $E_s(t)$ remains finite. If $p = 1$, i.e. if the nonlinear term in (3.4) is quadratic, then for both $n = 2$ and $n = 3$, the integral $\int_0^\infty (1+t)^{-\frac{(n-1)p}{2}} \, dt = \infty$, and we would not be able to conclude solutions exist for all time (for small initial data) from this analysis. In fact, if $p = 1$ and $n = 2$, the existence time we expect from this analysis would be of order$O(\epsilon^{-2})$ for data of size ϵ. On the other hand, if $p \ge 2$, i.e. if there is no quadratic nonlinearity, then we can expect to prove longer time existence for solutions of (3.4) starting from small initial data. In fact, we can expect for $n = 2$, $p = 2$, an existence time period of $[0, e^{c/\epsilon^2}]$; and for $n = 3$, $p = 2$, an existence time period $[0, \infty)$ for data of size ϵ, when ϵ is sufficiently small.

So the question now is whether there is another unknown $\mathfrak{v}$ that satisfies an equation of the type

$$\partial_t^2 \mathfrak{v} + |D|\mathfrak{v} = F_1(\mathfrak{v}_t, |D|\mathfrak{v}, \mathfrak{u}_t, |D|\mathfrak{u}) \tag{3.5}$$

with F_1 containing no quadratic nonlinearities and $\|\mathfrak{v}\| \approx \|\mathfrak{u}\|$ in various norms $\|\cdot\|$ involved in the course of analysis. The idea of finding such an unknown $\mathfrak{v}$ is the so called method of normal forms, originally introduced by Poincaré in solving ordinary differential equations. Certainly in most cases, one should not expect such a new unknown $\mathfrak{v}$ exist. For quadratic Klein-Gordon equation however, Shatah [**24**] succeeded in finding a bilinear normal form transformation of the type

$$v = u + B(u, u) \tag{3.6}$$

with $B(u, u)$ bilinear, canceling out the quadratic nonlinear terms in the Klein-Gordon equation, and satisfying the norm equivalence $\|u + B(u, u)\| \approx \|u\|$ for small $\|u\|$. However using the bilinear normal form ansatz $\mathfrak{v} = \mathfrak{u} + \mathfrak{B}(\mathfrak{u}, \mathfrak{u})$, we found that in order to cancel the quadratic terms in the water wave system (2.6)-(2.5) (or (5.21)-(5.22) of [**33**] for 3D water waves), the bilinear form $\mathfrak{B}(\mathfrak{u}, \mathfrak{u})$ must contain a small divisor of the type

$$\mathfrak{B}(\mathfrak{u}, \mathfrak{u}) = \iint e^{i\alpha\cdot(\eta_1+\eta_2)} m(\eta_1, \eta_2) \hat{\mathfrak{u}}(\eta_1)\hat{\mathfrak{u}}(\eta_2)\, d\eta_1\, d\eta_2 \tag{3.7}$$

where

$$\overline{\lim_{\eta_1 \to 0}} |\eta_1| m(\eta_1, \eta_2) \neq 0$$

Proving $\|\mathfrak{u} + \mathfrak{B}(\mathfrak{u}, \mathfrak{u})\| \approx \|\mathfrak{u}\|$ for small $\|\mathfrak{u}\|$ in such norms $\|\cdot\| = \|\cdot\|_{L^2(\mathbb{R}^{n-1})}$ is then impossible, as it amounts to proving such an inequality $\|f\|_{L^2(\mathbb{R}^{n-1})} \lesssim \|\nabla f\|_{L^2(\mathbb{R}^{n-1})}$.

After further analysis and reformulation of the water wave system (2.6)-(2.5) (or (5.21)-(5.22) of [**33**] for 3d water waves), we realized that the small divisor might be attributable to the particular coordinate system used in (2.6)-(2.5) (or (5.21)-(5.22) of [**33**] for 3d water waves), choosing an appropriate coordinate system (if it exists) might resolve this difficulty. Indeed through much further effort, we find new unknowns and new coordinate systems for both the two and three dimensional water waves, such that in these coordinate systems, the new unknown functions satisfy equations with no quadratic nonlinearities.

3.1. The transformation for the 2D water waves. We give the transformation we constructed for the 2D water waves, c.f. [**34**].

PROPOSITION 3.1 (Proposition 2.3 of [**34**]). *Let $\xi = \xi(\alpha, t)$ be a solution of the 2d water wave system* (2.2). *Let $\Pi = (I - \mathfrak{H})(\xi - \bar{\xi})$, $k = k(\cdot, t) : \mathbb{R} \to \mathbb{R}$ be a diffeomorphism. Let $\zeta = \xi \circ k^{-1}$, $D_t\zeta = \xi_t \circ k^{-1}$. We have*

(3.8)
$$(D_t^2 - iA\partial_\alpha)(\Pi \circ k^{-1}) = \frac{4}{\pi}\int \frac{(D_t\zeta(\alpha,t) - D_t\zeta(\beta,t))(\operatorname{Im}\zeta(\alpha,t) - \operatorname{Im}\zeta(\beta,t))}{|\zeta(\alpha,t) - \zeta(\beta,t)|^2}\partial_\beta D_t\zeta(\beta,t)d\beta + \frac{2}{\pi}\int \left(\frac{D_t\zeta(\alpha,t) - D_t\zeta(\beta,t)}{\zeta(\alpha,t) - \zeta(\beta,t)}\right)^2 \partial_\beta \operatorname{Im}\zeta(\beta,t)d\beta$$

Here $\operatorname{Im}\zeta$ *is the imaginary part of* ζ.

Notice that the right hand side of equation (3.8) is cubicly small if the velocity $D_t\zeta$, the height function $\operatorname{Im}\zeta$ of the interface (and their derivatives) are small. But the left hand side of (3.8) still contains quadratic nonlinearities. We resolved this difficulty by choosing an appropriate coordinate change k.

Let[2][3]

$$k(\alpha, t) = \bar{\xi}(\alpha, t) + \frac{1}{2}(I + \mathfrak{H})(I + \mathfrak{K})^{-1}(\xi - \bar{\xi}) \tag{3.9}$$

where $\mathfrak{K} = \operatorname{Re}\mathfrak{H}$ is the double layered potential operator. In Proposition 3.2 we will see that for the diffeomorphism k as defined by (3.9), the quantities $b = k_t \circ k^{-1}$ and $A - 1 = (\mathfrak{a}k_\alpha) \circ k^{-1} - 1$ are quadratically small provided the velocity $D_t\zeta$ and the height function $\operatorname{Im}\zeta$ of the interface (and their derivatives) are small. Therefore when the diffeomorephism k is given by (3.9), the left hand side of (3.8) is a linear term $(\partial_t^2 - i\partial_\alpha)(\Pi \circ k^{-1})$ plus some terms that are cubicly small.

PROPOSITION 3.2 (Proposition 2.4 of [**34**]). *Let k be as given in* (3.9), *$b = k_t \circ k^{-1}$ and $A = (\mathfrak{a}k_\alpha) \circ k^{-1}$. We have*

$$\begin{aligned}(I - \mathcal{H})b &= -[D_t\zeta, \mathcal{H}]\frac{\bar{\zeta}_\alpha - 1}{\zeta_\alpha} \\ (I - \mathcal{H})(A - 1) &= i[D_t\zeta, \mathcal{H}]\frac{\partial_\alpha \overline{D_t\zeta}}{\zeta_\alpha} + i[D_t^2\zeta, \mathcal{H}]\frac{\bar{\zeta}_\alpha - 1}{\zeta_\alpha}\end{aligned} \tag{3.10}$$

---

[2]The first coordinate change as defined in (2.18) of [**34**] used the Riemann mapping. However there is no Riemann mapping in 3D. To construct the coordinate change for the 3D water waves, we realized that the only property used of the Riemann mapping is that it is a holomorphic function in the fluid domain $\Omega(t)$ with its imaginary part equal to zero on $\Sigma(t)$. We therefore came up with a construction for the 3D water waves using this idea. The current formulation as in (3.9) also uses this idea and is used in the author's recent work with N. Totz on a rigorous justification of the modulation approximation for the 2D water wave problem [**31**].

[3]This $k = k(\cdot, t) : \mathbb{R} \to \mathbb{R}$ as defined in (3.9) is a diffeomorphism provided the interface $\xi = \xi(\cdot, t)$ is a small perturbation from the flat wave. This can be proved in a similar way as that in [**34**].

Therefore the quantity $\Pi \circ k^{-1} = (I - \mathcal{H})(\zeta - \bar{\zeta})$, with the coordinate change k given as in (3.9), satisfies an equation of the type $(\partial_t^2 - i\partial_\alpha)(\Pi \circ k^{-1}) = \mathcal{G}$ with $\mathcal{G}$ containing only nonlinear terms of cubic and higher orders. We make the following remarks:

REMARK 3.3.[4]

1. The transformation $I - \mathfrak{H}$ and the coordinate change k as given in (3.9) are fully nonlinear in terms of the unknown function ζ and its derivatives.
2. The bilinear part of the quantity $(I - \mathfrak{H})(\xi - \bar{\xi})$ has a bounded Fourier symbol. The coordinate change k has taken care of the difficulty of the small divisor in the Fourier symbol of the bilinear normal form transformation. This can be easily seen in the following calculation.

Assume that a normal form transformation $\mathfrak{G}$ is exactly obtained from the original unknown $\mathfrak{g}$ by a change of coordinates k^{-1}, i.e. $\mathfrak{G}(X) = \mathfrak{g}(k^{-1}(X))$, with $k(X) - X$ a small quantity depending on $\mathfrak{g}$. By the chain rule we know $\|\mathfrak{G}\| \approx \|\mathfrak{g}\|$ in various norms $\|\cdot\|$ provided the norms of $\nabla k^{-1} - I$ are finite, here I is the identity map. On the other hand, from the Taylor expansion

$$\mathfrak{G}(X) = \mathfrak{g}(k^{-1}(X)) = \mathfrak{g}(X) + \nabla\mathfrak{g}(X) \cdot (k^{-1}(X) - X) + \dots, \tag{3.11}$$

we know that if one only retains the linear and bilinear parts of the transformation $\mathfrak{g} \circ k^{-1}$, then one gets

$$\mathfrak{G}_1(X) = \mathfrak{g}(X) + \nabla\mathfrak{g}(X) \cdot (k^{-1}(X) - X)^5 \tag{3.12}$$

To obtain $\|\mathfrak{G}_1\| \approx \|\mathfrak{g}\|$ in various norms $\|\cdot\|$ for small $\|\mathfrak{g}\|$, one cannot avoid engaging the smallness of the norms of the primitive $k^{-1}(X) - X$, and the dependence on $\nabla\mathfrak{g}$ in the quadratic term can also be problematic. Therefore if one uses a full nonlinear transformation, in this case a full coordinate change, one doesn't encounter the problem of small divisors as one would if one considered a transformation consisting of only linear and bilinear forms of $\mathfrak{g}$.

3. Notice that the linear operator of the equation (3.8) is $\partial_t^2 - i\partial_\alpha$. The difference of $(\partial_t^2 + |D|)(\Pi \circ k^{-1})$ and $(\partial_t^2 - i\partial_\alpha)(\Pi \circ k^{-1})$ contains quadratic nonlinearities.

The proofs of Propositions 3.1, 3.2 use the following identities, c.f. [**34**]. Let $\xi = \xi(\cdot, t)$ define a non-selfintersecting curve, and $\mathfrak{H}$ be the Hilbert transform as defined in (2.1).

[4]For more detailed discussion of the normal form transformation of the water wave equations, we refer the reader to Appendix C of [**35**].

[5]This suggests an algorithm to handle problems where the bilinear normal form transformation contain small divisors attributable to a change of coordinates.

LEMMA 3.4 (Lemma 2.1 of [**34**]). *Assume that* $\xi_t, \xi_\alpha - 1 \in C^1([0,T], H^1(\mathbb{R}))$, $f \in C^1(\mathbb{R} \times (0,T))$ *satisfies* $f_\alpha(\alpha,t) \to 0$, *as* $|\alpha| \to \infty$. *We have*

(3.13)

$$[\partial_t, \mathfrak{H}]f = [\xi_t, \mathfrak{H}]\frac{f_\alpha}{\xi_\alpha}$$

$$[\partial_t^2, \mathfrak{H}]f = [\xi_{tt}, \mathfrak{H}]\frac{f_\alpha}{\xi_\alpha} + 2[\xi_t, \mathfrak{H}]\frac{f_{t\alpha}}{\xi_\alpha} - \frac{1}{\pi i}\int \left(\frac{\xi_t(\alpha,t) - \xi_t(\beta,t)}{\xi(\alpha,t) - \xi(\beta,t)}\right)^2 f_\beta \, d\beta$$

$$[\mathfrak{a}\partial_\alpha, \mathfrak{H}]f = [\mathfrak{a}\xi_\alpha, \mathfrak{H}]\frac{f_\alpha}{\xi_\alpha}, \quad \partial_\alpha \mathfrak{H} f = \xi_\alpha \mathfrak{H}\frac{f_\alpha}{\xi_\alpha}$$

$$[\partial_t^2 - i\mathfrak{a}\partial_\alpha, \mathfrak{H}]f = 2[\xi_t, \mathfrak{H}]\frac{f_{t\alpha}}{\xi_\alpha} - \frac{1}{\pi i}\int \left(\frac{\xi_t(\alpha,t) - \xi_t(\beta,t)}{\xi(\alpha,t) - \xi(\beta,t)}\right)^2 f_\beta \, d\beta$$

$$(I - \mathfrak{H})(-i\mathfrak{a}_t\bar{\xi}_\alpha) = 2[\xi_{tt}, \mathfrak{H}]\frac{\bar{\xi}_{t\alpha}}{\xi_\alpha} + 2[\xi_t, \mathfrak{H}]\frac{\bar{\xi}_{tt\alpha}}{\xi_\alpha} - \frac{1}{\pi i}\int \left(\frac{\xi_t(\alpha,t) - \xi_t(\beta,t)}{\xi(\alpha,t) - \xi(\beta,t)}\right)^2 \bar{\xi}_{t\beta} \, d\beta$$

The proof of Lemma 3.4 is straightforwardly integration by parts, we omit the proof.

Let Ω be a C^1 domain in the complex plane. Assume that the boundary of Ω is parametrized by $z = z(\alpha)$, $\alpha \in \mathbb{R}$, oriented in the clockwise sense, and there exist constants μ_1, $\mu_2 > 0$, such that $\mu_1|\alpha - \beta| \le |z(\alpha) - z(\beta)| \le \mu_2|\alpha - \beta|$ for all α, $\beta \in \mathbb{R}$. Let

$$\mathbf{H}f(\alpha) = \frac{1}{\pi i} p.v. \int \frac{z_\beta(\beta)}{z(\alpha) - z(\beta)} f(\beta) \, d\beta$$

be the associated Hilbert transform. We know $\mathbf{H}^2 = I$ in L^2, we use the convention $\mathbf{H}1 = 0$. We have

LEMMA 3.5 (Lemma 2.2 of [**34**]).

1. $f(\cdot) = F(z(\cdot)) \in L^2(\partial\Omega)$ *is the boundary value of a holomorphic function* F *in* Ω *if and only if* $f = \mathbf{H}f$.
2. *If* $f = \mathbf{H}f$, $g = \mathbf{H}g$, *then* $[f, \mathbf{H}]g = 0$.
3. *For any* $f, g \in L^2(\partial\Omega)$, *we have* $[f, \mathbf{H}]\mathbf{H}g = -[\mathbf{H}f, \mathbf{H}]g$.

The first statement in Lemma 3.5 is a consequence of the Cauchy integral formula. The second statement follows from the fact that the product of holomorphic functions is holomorphic. For details of the proof, see [**34**].

We now give the proofs for Propositions 3.1, 3.2.[6] We first prove (3.8). We have, for $\xi = \xi(\cdot,t)$ satisfying the water wave system (2.2),

$$\begin{aligned}(\partial_t^2 - i\mathfrak{a}\partial_\alpha)\{(I - \mathfrak{H})(\xi - \bar{\xi})\} = (I - \mathfrak{H})\{(\partial_t^2 - i\mathfrak{a}\partial_\alpha)(\xi - \bar{\xi})\} \\ - [\partial_t^2 - i\mathfrak{a}\partial_\alpha, \mathfrak{H}](\xi - \bar{\xi})\end{aligned}$$

[6]The proofs are taken from [**34**].

$$= (I - \mathfrak{H})(-2\bar{\xi}_{tt}) - 2[\xi_t, \mathfrak{H}]\frac{\xi_{t\alpha} - \bar{\xi}_{t\alpha}}{\xi_\alpha}$$

$$+ \frac{1}{\pi i}\int (\frac{\xi_t(\alpha,t) - \xi_t(\beta,t)}{\xi(\alpha,t) - \xi(\beta,t)})^2 \partial_\beta(\xi(\beta,t) - \bar{\xi}(\beta,t))\, d\beta$$

$$= -2[\xi_t, \mathfrak{H}]\frac{\xi_{t\alpha}}{\xi_\alpha} + \frac{1}{\pi i}\int (\frac{\xi_t(\alpha,t) - \xi_t(\beta,t)}{\xi(\alpha,t) - \xi(\beta,t)})^2 (\xi - \bar{\xi})_\beta\, d\beta$$

$$= -2[\xi_t, \mathfrak{H}\frac{1}{\xi_\alpha} + \bar{\mathfrak{H}}\frac{1}{\bar{\xi}_\alpha}]\xi_{t\alpha}$$

$$+ \frac{1}{\pi i}\int (\frac{\xi_t(\alpha,t) - \xi_t(\beta,t)}{\xi(\alpha,t) - \xi(\beta,t)})^2 (\xi_\beta - \bar{\xi}_\beta)\, d\beta \tag{3.14}$$

Here in the second and third steps we used (2.2) and Lemma 3.4, and in the fourth step we inserted the term $[\xi_t, \bar{\mathfrak{H}}\frac{1}{\bar{\xi}_\alpha}]\xi_{t\alpha}$ to make it an at least cubic nonlinear term. $[\xi_t, \bar{\mathfrak{H}}\frac{1}{\bar{\xi}_\alpha}]\xi_{t\alpha} = 0$ because both $\bar{\xi}_t$ and $\frac{\bar{\xi}_{t\alpha}}{\bar{\xi}_\alpha}$ are traces of some holomorphic functions in $\Omega(t)$ and because of the part 2 of Lemma 3.5. Applying the change of coordinates $U_{k^{-1}}$ to (3.14), and expanding the first term in the last equality, we arrive at (3.8).

Now for the proof of Proposition 3.2, we notice that for k be as given in (3.9),

$$k(\alpha,t) - \bar{\xi}(\alpha,t) = \frac{1}{2}(I + \mathfrak{H})(I + \mathfrak{K})^{-1}(\xi - \bar{\xi})$$

is the trace on $\Sigma(t)$ of a holomorphic function in $\Omega(t)$, so (3.10) can be proved in exactly the same way as that of Proposition 2.4 in [**34**]: Let Φ be the holomorphic function in $\Omega(t)$, such that

$$\bar{\xi}(\alpha,t) - k(\alpha,t) = \Phi(\xi(\alpha,t),t).$$

We have,

$$\bar{\xi}_t - k_t = \Phi_t + \Phi_\xi \xi_t, \quad \bar{\xi}_\alpha - k_\alpha = \Phi_\xi \xi_\alpha. \tag{3.15}$$

Using the fact Φ_t, Φ_ξ are holomorphic in $\Omega(t)$ and Lemma 3.5, we obtain

$$-(I - \mathfrak{H})k_t = (I - \mathfrak{H})(\bar{\xi}_t - k_t)$$

$$= (I - \mathfrak{H})\{\Phi_\xi \xi_t\} = [\xi_t, \mathfrak{H}]\Phi_\xi = [\xi_t, \mathfrak{H}]\frac{\bar{\xi}_\alpha - k_\alpha}{\xi_\alpha}. \tag{3.16}$$

Now using (3.15), (2.2), and the fact that Φ_ξ is holomorphic in $\Omega(t)$, we also have

$$(I - \mathfrak{H})(i\mathfrak{a}\bar{\xi}_\alpha - i\mathfrak{a}k_\alpha) = (I - \mathfrak{H})(i\mathfrak{a}\xi_\alpha\Phi_\xi) = [\xi_{tt}, \mathfrak{H}]\Phi_\xi$$

therefore using Lemma 3.4, (2.2) and (3.15),

$$-(I - \mathfrak{H})(i\mathfrak{a}k_\alpha) = -(I - \mathfrak{H})(i\mathfrak{a}\bar{\xi}_\alpha) + [\xi_{tt}, \mathfrak{H}]\Phi_\xi$$

$$= -i + [\xi_t, \mathfrak{H}]\frac{\bar{\xi}_{t\alpha}}{\xi_\alpha} + [\xi_{tt}, \mathfrak{H}]\frac{\bar{\xi}_\alpha - k_\alpha}{\xi_\alpha} \tag{3.17}$$

Applying the change of variable U_k^{-1} to (3.16), (3.17), we arrive at (3.10).

3.2. The transformation for the 3D water waves. We study the global in time behavior of the 3D water wave problem in the setting of the Clifford Algebra $\mathcal{C}(V_2)$. We further introduce the following notations.

We know an element $\sigma \in \mathcal{C}(V_2)$ has a unique representation $\sigma = \sigma_0 + \sum_{i=1}^3 \sigma_i e_i$, with $\sigma_i \in \mathbb{R}$ for $0 \le i \le 3$; We call σ a vector if $\sigma_0 = 0$. If not specified, we always assume in such an expression $\sigma = \sigma_0 + \sum_{i=1}^3 \sigma_i e_i$ that $\sigma_i \in \mathbb{R}$ for $0 \le i \le 3$. We define $\overline{\sigma} = e_3 \sigma e_3$, the conjugate of σ. We identify a point or a vector $\xi = (x, y, z) \in \mathbb{R}^3$ with its $\mathcal{C}(V_2)$ counterpart $\xi = xe_1 + ye_2 + ze_3$. For vectors $\xi,\ \eta \in \mathcal{C}(V_2)$, we know

$$\xi\eta = -\xi \cdot \eta + \xi \times \eta, \tag{3.18}$$

where $\xi \cdot \eta$ is the dot product, $\xi \times \eta$ the cross product. For vectors $\xi,\ \zeta,\ \eta$, $\xi(\zeta \times \eta)$ is obtained by first finding the cross product $\zeta \times \eta$, then regard it as a Clifford vector and calculating its multiplication with ξ by the rule (2.7). We write $\nabla = (\partial_x, \partial_y, \partial_z)$. We abbreviate notations such as

$$\begin{aligned}\mathfrak{H}_\Sigma f(\alpha, \beta) &= \iint K(\eta(\alpha', \beta') - \eta(\alpha, \beta))\,(\eta'_{\alpha'} \times \eta'_{\beta'})\, f(\alpha', \beta')\, d\alpha' d\beta' \\ &= \iint K(\eta' - \eta)\,(\eta'_{\alpha'} \times \eta'_{\beta'})\, f'\, d\alpha' d\beta' = \iint K\, N'\, f'\, d\alpha' d\beta'.\end{aligned}$$

Similar to the 2D case, we also have $\mathfrak{H}_\Sigma^2 = I$ in L^2; we use the convention $\mathfrak{H}_\Sigma 1 = 0$.

We now give the transformation for the 3D water wave equation. Let $\Omega(t)$ be the fluid region, $\Sigma(t)$ be the fluid interface in $\mathbb{R}^3$. Let $\Sigma(t)$: $\xi(\alpha, \beta, t) = x(\alpha, \beta, t)e_1 + y(\alpha, \beta, t)e_2 + z(\alpha, \beta, t)e_3$, $(\alpha, \beta) \in \mathbb{R}^2$ be the parameterization of the interface $\Sigma(t)$ at time t in Lagrangian coordinates (α, β) with $N = \xi_\alpha \times \xi_\beta$ pointing out of the fluid domain $\Omega(t)$. Let $\mathfrak{H} = \mathfrak{H}_{\Sigma(t)}$ be defined as in (2.8), and

$$\mathfrak{a} = -\frac{1}{|N|}\frac{\partial P}{\partial \mathbf{n}}.$$

We know from [**33**] that $\mathfrak{a} > 0$ and equation (1.1) ($n = 3$) is equivalent to the following nonlinear system defined on the interface $\Sigma(t)$:

$$\begin{cases} \xi_{tt} + e_3 = \mathfrak{a} N \\ \xi_t = \mathfrak{H} \xi_t \end{cases} \tag{3.19}$$

Let $U_g f(\alpha, \beta, t) = f(g(\alpha, \beta, t), t) = f \circ g(\alpha, \beta, t)$. For fixed t, let $k = k(\cdot, t) = k_1 e_1 + k_2 e_2 : \mathbb{R}^2 \to \mathbb{R}^2$ be a diffeomorphism with its Jacobian $J(k(t)) > 0$. Let k^{-1} be such that $k \circ k^{-1}(\alpha, \beta, t) = \alpha e_1 + \beta e_2$. Define

$$\zeta = \xi \circ k^{-1}, \quad b = k_t \circ k^{-1}, \quad A \circ k e_3 = \mathfrak{a} J(k) e_3 = \mathfrak{a} k_\alpha \times k_\beta \tag{3.20}$$

Let $D_t = U_k^{-1}\partial_t U_k$ be the material derivative, $\mathcal{N} = \zeta_\alpha \times \zeta_\beta$. By a simple application of the chain rule, we know

(3.21) $D_t = \partial_t + b\cdot(\partial_\alpha, \partial_\beta), \quad U_k^{-1}(\mathfrak{a}N\times\nabla)U_k = A\mathcal{N}\times\nabla = A(\zeta_\beta\partial_\alpha - \zeta_\alpha\partial_\beta),$

and $U_k^{-1}\mathfrak{H}U_k = \mathcal{H}$, with

(3.22)
$$\mathcal{H}f(\alpha,\beta,t) = \iint K(\zeta(\alpha',\beta',t) - \zeta(\alpha,\beta,t))(\zeta'_{\alpha'}\times\zeta'_{\beta'})f(\alpha',\beta',t)\,d\alpha'\,d\beta'$$

We have

PROPOSITION 3.6 (Proposition 1.3 of [**35**]). *Let $\xi = \xi(\alpha,\beta,t)$ be a solution of the 3d water wave system (3.19). Let $\Pi = (I - \mathfrak{H})(\xi - \bar{\xi})$, and for fixed t, $k(\cdot,t) : \mathbb{R}^2 \to \mathbb{R}^2$ be a diffeomorphism. We have*

(3.23)
$$\begin{aligned}
&(D_t^2 - A\mathcal{N}\times\nabla)(\Pi\circ k^{-1})\\
&\quad = 2\iint K(\zeta'-\zeta)\,(D_t\zeta - D'_t\zeta')\times(\zeta'_{\beta'}\partial_{\alpha'} - \zeta'_{\alpha'}\partial_{\beta'})\overline{D'_t\zeta'}\,d\alpha' d\beta'\\
&\quad\quad - \iint K(\zeta'-\zeta)\,(D_t\zeta - D'_t\zeta')\times((D'_t\zeta')_{\beta'}\partial_{\alpha'}\\
&\quad\quad - (D'_t\zeta')_{\alpha'}\partial_{\beta'})(\zeta' - \bar{\zeta}')\,d\alpha' d\beta'\\
&\quad\quad - \iint D_t K(\zeta'-\zeta)\,(D_t\zeta - D'_t\zeta')\times(\zeta'_{\beta'}\partial_{\alpha'} - \zeta'_{\alpha'}\partial_{\beta'})(\zeta' - \bar{\zeta}')\,d\alpha' d\beta'
\end{aligned}$$

We see that the second and third terms in the right hand side of (3.23) are cubicly small provided the velocity $D_t\zeta$ and the steepness of the height function $\partial_\alpha(\zeta-\bar{\zeta})$, $\partial_\beta(\zeta-\bar{\zeta})$ are small, while the first term appears to be only quadratically small. Unlike the 2D case, multiplications of Clifford analytic functions are not necessarily analytic, so we cannot reduce the first term at the right hand side of equation (3.23) into a cubic form. However we notice that $\overline{D_t\zeta}$ is almost analytic[7] in the air region $\Omega(t)^c$, and this implies that the first term is almost analytic in the fluid domain $\Omega(t)$. Notice that the left hand side of (3.23) is almost analytic in the air region. The orthogonality of the projections $(I - \mathcal{H})$ and $(I + \mathcal{H})$ allows us to reduce the first term to cubic in energy estimates.

Now the left hand side of (3.23) still contains quadratic terms. Similar to the 2D case, this difficulty is resolved by choosing an appropriate coordinate change k. Let

(3.24) $$k = k(\alpha,\beta,t) = \xi(\alpha,\beta,t) - (I + \mathfrak{H})z(\alpha,\beta,t)e_3 + \mathfrak{K}z(\alpha,\beta,t)e_3$$

[7]Since the order of smallness is what matters, here a quantity X of size $O(\epsilon^N)$, in other words of order N, is said to be almost analytic in the fluid region $\Omega(t)$ (respectively in the air region $\Omega(t)^c$), if $(I - \mathcal{H})X$ (respectively $(I + \mathcal{H})X$) is at most of size $O(\epsilon^{N+1})$, or in other words at least of order $N + 1$.

Here $\mathfrak{K} = \operatorname{Re}\mathfrak{H}$:

$$\mathfrak{K}f(\alpha,\beta,t) = -\iint K(\xi(\alpha',\beta',t)-\xi(\alpha,\beta,t))\cdot N' f(\alpha',\beta',t)\,d\alpha'\,d\beta' \tag{3.25}$$

is the double layered potential operator. It is clear that the e_3 component of k as defined in (3.24) is zero. In fact, the real part of k is also zero. This is because

$$\begin{aligned}
&\iint K(\xi'-\xi)\times(\xi'_{\alpha'}\times\xi'_{\beta'})z'e_3\,d\alpha'\,d\beta'\\
&=\iint(\xi'_{\alpha'}\xi'_{\beta'}\cdot K-\xi'_{\beta'}\xi'_{\alpha'}\cdot K)z'e_3\,d\alpha'\,d\beta'\\
&=-2\iint(\xi'_{\alpha'}\partial_{\beta'}\Gamma(\xi'-\xi)-\xi'_{\beta'}\partial_{\alpha'}\Gamma(\xi'-\xi))z'e_3\,d\alpha'\,d\beta'\\
&=2\iint\Gamma(\xi'-\xi)(\xi'_{\alpha'}z_{\beta'}-\xi'_{\beta'}z_{\alpha'})e_3\,d\alpha'\,d\beta'\\
&=2\iint\Gamma(\xi'-\xi)(N'_1e_1+N'_2e_2)\,d\alpha'\,d\beta'
\end{aligned}$$

So

$$\mathfrak{H}ze_3=\mathfrak{K}ze_3+2\iint\Gamma(\xi'-\xi)(N'_1e_1+N'_2e_2)\,d\alpha'\,d\beta' \tag{3.26}$$

This shows that the mapping k defined in (3.24) has only the e_1 and e_2 components $k=(k_1,k_2)=k_1e_1+k_2e_2$. If $\Sigma(t)$ is a graph with small steepness, i.e. if z_α and z_β are small, then the Jacobian of $k=k(\cdot,t)$: $J(k)=J(k(t))=\partial_\alpha k_1\partial_\beta k_2-\partial_\alpha k_2\partial_\beta k_1>0$ and $k(\cdot,t):\mathbb{R}^2\to\mathbb{R}^2$ defines a valid coordinate change (c.f. [**35**]).

From the following proposition we see that if k is as given in (3.24), then b, $A-1$ are consisting of only quadratic and higher order terms. Let

$$\mathcal{K}=\operatorname{Re}\mathcal{H}=U_k^{-1}\mathfrak{K}U_k,\quad P=\alpha e_1+\beta e_2,\quad\text{and}\quad\zeta=P+\lambda. \tag{3.27}$$

PROPOSITION 3.7 (Proposition 1.4 of [**35**]). *Let k be as given in (3.24). Let $b=k_t\circ k^{-1}$ and $A\circ k=\mathfrak{a}J(k)$. We have*

$$\begin{aligned}
b&=\frac12(\mathcal{H}-\overline{\mathcal{H}})\overline{D_t\zeta}-\frac12[D_t,\mathcal{H}-\mathcal{K}](\zeta-\bar\zeta)+\frac12\mathcal{K}(D_t\zeta-\overline{D_t\zeta})\\
(A-1)e_3&=\frac12(-\mathcal{H}+\overline{\mathcal{H}})\overline{D_t^2\zeta}+\frac12([D_t,\mathcal{H}]D_t\zeta-\overline{[D_t,\mathcal{H}]D_t\zeta})\\
&\quad+\frac12[A\mathcal{N}\times\nabla,\mathcal{H}](\zeta-\bar\zeta)-\frac12A\zeta_\beta\times(\partial_\alpha\mathcal{K}(\zeta-\bar\zeta))\\
&\quad+\frac12A\zeta_\alpha\times(\partial_\beta\mathcal{K}(\zeta-\bar\zeta))+A\partial_\alpha\lambda\times\partial_\beta\lambda
\end{aligned} \tag{3.28}$$

Here $\overline{\mathcal{H}}=e_3\mathcal{H}e_3$.

Let $\chi = \Pi \circ k^{-1}$ with k be as defined by (3.24). Therefore the left hand side of equation (3.23) is

$$(\partial_t^2 - e_2\partial_\alpha + e_1\partial_\beta)\chi - \partial_\beta\lambda\partial_\alpha\chi + \partial_\alpha\lambda\partial_\beta\chi + \text{cubic and higher order terms}$$

The quadratic term $\partial_\beta\lambda\partial_\alpha\chi - \partial_\alpha\lambda\partial_\beta\chi$ is new in 3D. We notice that this is one of the null forms studied in [**17**] and we find that it is also null for our equation and can be written as the factor $1/t$ times a quadratic expression involving some "invariant vector fields" for $\partial_t^2 - e_2\partial_\alpha + e_1\partial_\beta$. Therefore this term is cubic in nature and equation (3.23) is of the type "linear + cubic and higher order perturbations".

3.3. Global in time behavior of solutions for the 2D and 3D water waves. With equations (3.8) and (3.23) available now we can study the global in time behavior of small solutions of the 2D and 3D water wave equations. We point out that it is more natural to treat $D_t^2 - iA\partial_\alpha$ and $D_t^2 - A\mathcal{N} \times \nabla$ as the main operators for the 2D and 3D water wave equations than treating them as perturbations of the linear operators $\partial_t^2 - i\partial_\alpha$ and $\partial_t^2 - e_2\partial_\alpha + e_1\partial_\beta$. The basic idea is what we illustrated with the model equation (3.4); we make this idea rigorous by the method of invariant vector fields. This involves constructing invariant vector fields for the operator $\partial_t^2 - e_2\partial_\alpha + e_1\partial_\beta$ (the invariant vector fields for $\partial_t^2 - i\partial_\alpha$ for the 2D water wave are available due to the well studied Schrödinger operator $i\partial_t - \partial_x^2$); proving generalized Sobolev inequalities that give $L^2 \to L^\infty$ estimates with the decay rate $1/t^{1/2}$ for the 2D and $1/t$ for the 3D water waves respectively; using equations (3.8) and (3.23) to show that properly constructed energies that involve invariant vector fields remain bounded for the time period $[0, e^{c/\epsilon}]$ for the 2D and for all time for the 3D water waves for data of size $O(\epsilon)$. We point out that not only does the projection $(I - \mathfrak{H})$ give us the quantity $(I - \mathfrak{H})(\xi - \bar{\xi})$, but it is also used in various ways to project away "quadratic noises" in the course of deriving the energy estimates. The almost global well-posedness for the 2D and global well-posedness for the 3D water wave equations follow from the local well-posedness results, the uniform boundedness of the energies and continuity arguments. We state our main theorems.

Let $|D| = \sqrt{-\Delta}$, $H^s(\mathbb{R}^{n-1}) = \{f \,|\, (I + |D|)^s f \in L^2(\mathbb{R}^{n-1})\}$, with $\|f\|_{H^s(\mathbb{R}^{n-1})} = \|(I + |D|)^s f\|_{L^2(\mathbb{R}^{n-1})}$.

2D water waves. Let $s \geq 12$, $\max\{[\frac{s}{2}] + 3, 11\} \leq l \leq s - 1$. Assume initially that

$$\begin{aligned} &\xi(\alpha, 0) = (\alpha, y(\alpha)), \quad \xi_t(\alpha, 0) = \mathfrak{u}(\alpha), \quad \xi_{tt}(\alpha, 0) = \mathfrak{w}(\alpha) \quad \alpha \in \mathbb{R}, \\ &\qquad\qquad\qquad\qquad\qquad \mathbf{v}(z, 0) = \mathbf{g}(z), \quad z \in \Omega(0) \end{aligned} \tag{3.29}$$

and the data in (3.29) satisfies the 2D water wave system (2.2). In particular $\bar{\mathbf{g}}$ is a holomorphic function in the initial fluid domain $\Omega(0)$ and

$\mathbf{g}(\xi(\alpha,0)) = \mathfrak{u}(\alpha)$. Let $\Gamma = \partial_\alpha, \alpha\partial_\alpha$. Assume that

$$\sum_{|j|\le s-1} (\|\Gamma^j y_\alpha\|_{H^{1/2}(\mathbb{R})} + \|\Gamma^j \mathfrak{u}\|_{H^{3/2}(\mathbb{R})} + \|\Gamma^j \mathfrak{w}\|_{H^1(\mathbb{R})}) < \infty.$$

Let

$$\epsilon = \sum_{|j|\le l} (\|\Gamma^j y\|_{H^1(\mathbb{R})} + \|\Gamma^j \mathfrak{u}\|_{H^1(\mathbb{R})}) + \sum_{j\le l-2} \|(z\partial_z)^j \bar{\mathbf{g}}\|_{L^2(\Omega(0))}.$$

THEOREM 3.8 (2D Theorem [**34**]). *There exist ϵ_0 and $c > 0$, such that for $\epsilon < \epsilon_0$, the initial value problem (2.2)–(3.29) has a unique classical solution for the time period $[0, e^{c/\epsilon}]$. During this time period, the interface is a graph, the solution is as regular as the initial data and remains small. Moreover the L^∞ norm of the steepness of the interface $\partial_\alpha(\xi - \bar{\xi})$, the velocity ξ_t and acceleration ξ_{tt} on the interface decay at rate $1/t^{1/2}$.*

3D water waves. Let $s \ge 27$, $\max\{[\frac{s}{2}] + 1, 17\} \le l \le s - 10$. Assume that initially

$$\xi(\alpha,\beta,0) = (\alpha,\beta,z^0(\alpha,\beta)), \quad \xi_t(\alpha,\beta,0) = \mathfrak{u}^0(\alpha,\beta),$$
$$\xi_{tt}(\alpha,\beta,0) = \mathfrak{w}^0(\alpha,\beta), \tag{3.30}$$

and the data in (3.30) satisfies the 3D water wave system (3.19). Let $\Gamma = \partial_\alpha,\ \partial_\beta,\ \alpha\partial_\alpha + \beta\partial_\beta,\ \alpha\partial_\beta - \beta\partial_\alpha$. Assume that

$$\sum_{\substack{|j|\le s-1\\ \partial=\partial_\alpha,\partial_\beta}} \|\Gamma^j |D|^{1/2} z^0\|_{L^2(\mathbb{R}^2)} + \|\Gamma^j \partial z^0\|_{H^{1/2}(\mathbb{R}^2)}$$
$$+ \|\Gamma^j \mathfrak{u}^0\|_{H^{3/2}(\mathbb{R}^2)} + \|\Gamma^j \mathfrak{w}^0\|_{H^1(\mathbb{R}^2)} < \infty \tag{3.31}$$

Let

$$\epsilon = \sum_{\substack{|j|\le l+3\\ \partial=\partial_\alpha,\partial_\beta}} \|\Gamma^j |D|^{1/2} z^0\|_{L^2(\mathbb{R}^2)} + \|\Gamma^j \partial z^0\|_{L^2(\mathbb{R}^2)}$$
$$+ \|\Gamma^j \mathfrak{u}^0\|_{H^{1/2}(\mathbb{R}^2)} + \|\Gamma^j \mathfrak{w}^0\|_{L^2(\mathbb{R}^2)}. \tag{3.32}$$

THEOREM 3.9 (3D Theorem, c.f. [**35**]). *There exists $\epsilon_0 > 0$, such that for $0 \le \epsilon \le \epsilon_0$, the initial value problem (3.19)–(3.30) has a unique classical solution globally in time. For each time $0 \le t < \infty$, the interface is a graph, the solution has the same regularity as the initial data and remains small. Moreover the L^∞ norm of the steepness and the acceleration on the interface, the derivative of the velocity on the interface decay at rate $\frac{1}{t}$.*

REMARK 3.10. In [**34, 35**] we considered only the case that the velocity $\mathbf{v} \to 0$ at the spatial infinity. One can also treat water waves with $\mathbf{v} \to c = (c', 0)$ a constant velocity at spatial infinity. Exact analogous computations lead to analogues of (3.8) and (3.23). This yields results like Theorems 3.8, 3.9 with $\mathbf{v} \to c$ at spatial infinity.

REMARK 3.11. Theorems 3.8, 3.9 and their analogues for $\mathbf{v} \to c$ at spatial infinity show that there are lower bounds on the size of possible solitary waves in 2D and 3D since the L^∞ decay of small solutions rules out small solitary waves.[8]

REMARK 3.12. Using the advantageous structure of the equation (3.8), N. Totz and the author obtained a rigorous justification of the modulation approximation to the 2D water wave system (1.1) ($n = 2$), c.f. [**31**].

References

[1] D. Ambrose, N. Masmoudi *The zero surface tension limit of two-dimensional water waves.* Comm. Pure Appl. Math. 58 (2005), no. 10, 1287–1315.

[2] T. Beale, T. Hou & J. Lowengrub *Growth rates for the linearized motion of fluid interfaces away from equilibrium* Comm. Pure Appl. Math. 46 (1993), no.9, 1269–1301.

[3] G. Birkhoff *Helmholtz and Taylor instability* Proc. Symp. in Appl. Math. Vol. XIII, pp.55–76.

[4] S. Chen & Y. Zhou *Decay rate of solutions to hyperbolic system of first order* Acta. Math. Sinica, English series. 1999, Vol. 15, no. 4, p.471–484.

[5] D. Christodoulou, H. Lindblad *On the motion of the free surface of a liquid* Comm. Pure Appl. Math. 53 (2000) no. 12, 1536–1602.

[6] D. Coutand, S. Shkoller *Wellposedness of the free-surface incompressible Euler equations with or without surface tension* J. AMS. 20 (2007), no. 3, 829–930.

[7] R.R. Coifman, Y. Meyer *Au delá des opérateurs pseudodifférentials* Asterisque 57, Societé Math. de France, 1978.

[8] R. Coifman, A. McIntosh and Y. Meyer *Líntegrale de Cauchy definit un operateur borne sur L^2 pour les courbes lipschitziennes*, Annals of Math, 116 (1982), 361–387.

[9] R. Coifman, G. David and Y. Meyer *La solution des conjectures de Calderón* Adv. in Math. 48, 1983, pp. 144–148.

[10] W. Craig *Non-existence of solitary water waves in three dimensions* R. Soc. Lond. Philos. Trans. Ser. A Math. Phys. Eng. Sci. 360 (2002), no. 1799, 2127–2135.

[11] W. Craig *An existence theory for water waves and the Boussinesq and Korteweg-devries scaling limits* Comm. in P. D. E. 10(8), 1985 pp.787–1003.

[12] C.D. Crapper *Introduction to water waves* Ellis Horwood Ltd. 1984.

[13] P. Germain, N. Masmoudi, & J. Shatah *Global solutions of the gravity water wave equation in dimension 3* Preprint July 2009.

[14] J. Gilbert & M. Murray *Clifford algebras and Dirac operators in harmonic analysis* Cambridge University Press, 1991.

[15] T. Iguchi *Well-posedness of the initial value problem for capillary-gravity waves* Funkcial. Ekvac. 44 (2001) no. 2, 219–241.

[16] C. Kenig *Elliptic boundary value problems on Lipschitz domains* Beijing Lectures in Harmonic Analysis, ed. by E. M. Stein, Princeton Univ. Press, 1986, p. 131–183.

[17] S. Klainerman *The null condition and global existence to nonlinear wave equations*, Lectures in Appl. Math. vol. 23 (1986), 293–325.

[18] D. Lannes *Well-posedness of the water-wave equations* J. Amer. Math. Soc. 18 (2005), 605–654.

[19] H. Lindblad *Well-posedness for the motion of an incompressible liquid with free surface boundary* Ann. of Math. 162 (2005), no. 1, 109–194.

[8]Non-existence of positively or negatively elevated solitary waves for 2-D infinite depth water wave with or without surface tension was established in [**29**]; for 3-D finite or infinite depth positively elevated water wave without surface tension it was established in [**10**].

[20] V. I. Nalimov *The Cauchy-Poisson Problem* (in Russian), Dynamika Splosh. Sredy 18, 1974, pp. 104–210.
[21] M. Ogawa, A. Tani *Free boundary problem for an incompressible ideal fluid with surface tension* Math. Models Methods Appl. Sci. 12, (2002), no. 12, 1725–1740.
[22] C. Sogge *Lectures on nonlinear wave equations* International Press, 1995.
[23] G. Schneider and E. Wayne *The long wave limit for the water wave problem I. The case of zero surface tension* Comm. Pure. Appl. Math. 53, 2000, no. 12, 1475–1535.
[24] J. Shatah *Normal forms and quadratic nonlinear Klein-Gordon equations* Comm. Pure Appl. Math. 38 (1985), 685–696.
[25] J. Shatah, C. Zeng *Geometry and a priori estimates for free boundary problems of the Euler's equation* Comm. Pure Appl. Math. V. 61. no.5 (2008) pp. 698–744
[26] E.M.Stein *Singular integrals and differentiability properties of functions* Princeton University Press, 1970.
[27] W. Strauss *Nonlinear wave equations* CBMS No.73 AMS, 1989.
[28] Sulem, C.; Sulem, P.-L., *The nonlinear Schrodinger equation. Self-focusing and wave collapse* Applied Math- ematical Sciences, 139. Springer-Verlag, New York, 1999.
[29] Sun, S. M. *Some analytical properties of capillary-gravity waves in two-fluid flows of infinite depth* Proc. Roy. Soc. Lond. Ser.A 453 (1997) no. 1961, 1153–1175.
[30] G. I. Taylor *The instability of liquid surfaces when accelerated in a direction perpendicular to their planes I.* Proc. Roy. Soc. London A 201, 1950, 192–196.
[31] N. Totz & S. Wu *A Rigorous justification of the modulation approximation to the 2D full water wave problem,* accepted by Comm. Math. Phys.
[32] S. Wu *Well-posedness in Sobolev spaces of the full water wave problem in 2-D* Inventiones Mathematicae 130, 1997, pp. 39–72.
[33] S. Wu *Well-posedness in Sobolev spaces of the full water wave problem in 3-D* Journal of the AMS. 12. no.2 (1999), pp. 445–495.
[34] S. Wu *Almost global wellposedness of the 2-D full water wave problem* Inventiones Mathematicae, Vol. 177 no.1 July 2009. pp. 45–135.
[35] S. Wu *Global well-posedness of the 3-D full water wave problem,* Inventiones Mathematicae, Vol. 184 no.1 April 2011, pp. 125–220.
[36] H. Yosihara *Gravity waves on the free surface of an incompressible perfect fluid of finite depth,* RIMS Kyoto 18, 1982, pp. 49–96.
[37] P. Zhang, Z. Zhang *On the free boundary problem of 3-D incompressible Euler equations.* Comm. Pure. Appl. Math. V. 61. no.7 (2008), pp. 877–940.

UNIVERSITY OF MICHIGAN, ANN ARBOR, MI

Made in United States
North Haven, CT
05 November 2026

21764585R00102